AF615641

CAMBRIDGE MONOGRAPHS ON MATHEMATICAL PHYSICS



# SEMI-CLASSICAL METHODS FOR NUCLEUS–NUCLEUS SCATTERING

# SEMI-CLASSICAL METHODS FOR NUCLEUS–NUCLEUS SCATTERING

**D. M. BRINK**

*Department of Theoretical Physics*
*University of Oxford*

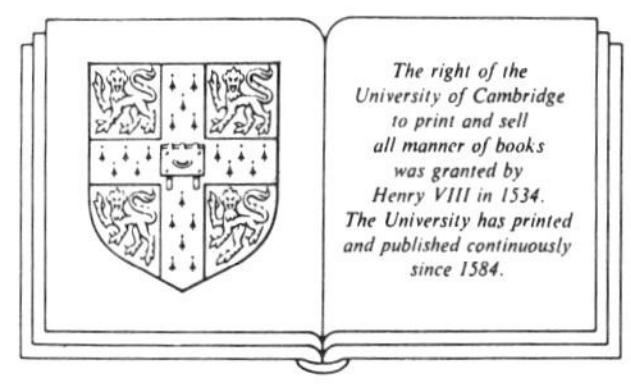


CAMBRIDGE UNIVERSITY PRESS

*Cambridge*

*London New York New Rochelle*
*Melbourne Sydney*

Published by the Press Syndicate of the University of Cambridge
The Pitt Building, Trumpington Street, Cambridge CB2 1RP
32 East 57th Street, New York, NY 10022, USA
10 Stamford Road, Oakleigh, Melbourne 3166, Australia



First published 1985

Printed in Great Britain at the University Press, Cambridge

Library of Congress catalogue card number: 84–23173

*British Library cataloguing in publication data*

Brink, D. M.
Semi-classical methods for nucleus–nucleus scattering – (Cambridge monographs on mathematical physics)
1. Nuclear reactions
I. Title
539.7′5 QC794

ISBN 0 521 23940 0

TM

# Contents

## APPENDICES

# Preface

Semi-classical methods are as old as quantum mechanics and an enormous amount of work has been done in the field. The *Wentzel, Kramers & Brillouin* (WKB) method was developed in 1926, immediately after the advent of wave mechanics. Feynman's path integral formulation of quantum mechanics gives an alternative approach to semi-classical methods. Indeed, the path integral theory emphasizes the close connection between classical mechanics and quantum mechanics.

Applications of semi-classical methods to nuclear reaction theory were initiated by the work of Blair on alpha-particle scattering and of Frahn and Venter on heavy-ion scattering. In the late 1960s papers began to appear basing semi-classical methods on the path integral approach. During the last decade there has been a period of intense activity in developing semi-classical methods for heavy-ion reactions. There has been cross fertilization from atomic physics and elementary particle physics. In this period a number of different schools have grown up, each one emphasizing different aspects of the theory and using different techniques. The aim of this book is to present an overall view of these lines of development and to attempt to draw them together.

The book is intended for advanced students and research workers interested in semi-classical methods applied to nuclear scattering problems. It is self-contained and there are introductory sections on the WKB approximation and the path integral method. Various formulae and descriptions of approximation methods are collected in appendices.

Many colleagues in different laboratories have helped me to clarify my ideas. I shall not try to name them all, but I have especially benefited from discussions with R. Anni, F. dos Aidos, C. Marty, N. Rowley, U. Smilansky and C.V. Sukumar.

I would like to express my thanks to J. S. Elkington of the Nuclear Physics Laboratory, Oxford, and to my son, T. D. Brink, for help with preparing the diagrams.

Finally, I want to thank my wife, Verena, not only for her patience and understanding, but also for her help in typing the manuscript and preparing the material for the publisher.

Oxford, May 1984 D. M. Brink

# 1
# Introduction

This book is concerned with the application of semi-classical methods to nuclear reaction theory. These methods are especially appropriate for the study of heavy-ion reactions when one nucleus is scattered by another, but they have also been used for alpha-particle scattering. Similar methods have also been applied to atomic and molecular collisions. Nuclear collisions are different in that the Coulomb force between the nuclei has a very important influence on the characteristics of the scattering.

A wide range of processes can occur in a heavy-ion collision. The simplest is elastic scattering, such as

$$^{16}\mathrm{O} + {}^{208}\mathrm{Pb} \rightarrow {}^{16}\mathrm{O} + {}^{208}\mathrm{Pb}$$

The differential cross-section and the total reaction cross-section can both be measured as a function of energy. Inelastic scattering is another possible process when one or both of the final nuclei is left in an excited state. When the final states have non-zero angular momentum their spins may be polarized. Differential cross-sections and final state polarizations can be measured in principle, although there may be practical difficulties.

Quasi-elastic transfer reactions occur when one or several nucleons are transferred between the target and projectile. An example is

$$^{12}\mathrm{C} + {}^{208}\mathrm{Pb} \rightarrow {}^{11}\mathrm{B} + {}^{209}\mathrm{Bi}$$

where a proton is transferred from the projectile into the target. It is often possible to distinguish individual final states in simple transfer reactions. When many nucleons are exchanged and much energy is transferred from relative motion into internal excitation then experiments can only measure cross-sections and energy losses summed over many final states. Such processes are called deep inelastic collisions.

In a fusion reaction the two nuclei combine to form a composite system. The final nucleus is produced in a highly excited state and decays by emitting nucleons, heavier particles or $\gamma$-rays. The final product observed in the experiment is called an evaporation residue if its mass is near that of the compound system. When most of the projectile attaches itself to the target and some smaller fragment is released immediately the process is termed 'incomplete fusion'.

The above reactions are the important ones for projectiles up to 10–15 MeV per nucleon. At higher energies a large part of the projectile and target may dissociate into smaller fragments. Such fragmentation processes are dominant for reactions initiated by relativistic projectiles.

In a collision between two nuclei the wave number $k$ of relative motion is $k = \mu v/\hbar$ where $\mu$ is the reduced mass and $v$ is the relative velocity. Usually the de Broglie wave length is small compared with some characteristic dimension of the system and the collision is close to being classical. Then semi-classical or quasi-classical methods can be used to describe the collision. This book reviews the semi-classical methods that have been developed for discussing nucleus–nucleus collisions.

The reduced mass of two nuclei with mass numbers $A_1, A_2$ can be estimated quite accurately by the formula

$$\mu = mA_1A_2/(A_1 + A_2)$$

where $m = 931.50\,\mathrm{MeV}/c^2$ is one atomic mass unit, a typical mass of a nucleon bound in a nucleus. The asymptotic relative velocity $v$ of the target and projectile is given by

$$v/c = (2E_{\mathrm{lab}}/(A_1mc^2))^{\frac{1}{2}} = 0.04634(E_{\mathrm{lab}}/A_1)^{\frac{1}{2}}$$

if $E_{\mathrm{lab}}$ is measured in MeV, and the wave number of relative motion is

$$k = \mu v/\hbar = 4.7206A_1A_2/(A_1 + A_2) \times v/c\,\mathrm{fm}^{-1}$$

Two nuclei interact strongly in a nucleus–nucleus collision if their distance of closest approach is less than a strong absorption radius $R_s$. A first estimate of $R_s$ would be the sum of the radii of the two nuclei, but in fact $R_s$ should be somewhat larger than this because of the surface diffuseness of the nuclear density distributions and the finite range of nuclear forces. An empirical estimate of $R_s$ which is suitable for qualitative considerations is

$$R_s \approx 1.5(A_1^{\frac{1}{3}} + A_2^{\frac{1}{3}})\,\mathrm{fm} \tag{1.1}$$

Nucleus–nucleus collisions differ from atomic or molecular collisions in that the nuclei have large charges and Coulomb forces are very important. The Coulomb length parameter $a_C$ is a characteristic distance for collisions between charged objects. In the case of two colliding nuclei with charge numbers $Z_1$ and $Z_2$ ($\mathrm{e}^2 = 1.43996\,\mathrm{MeV\,fm}$)

$$a_C = Z_1Z_2e^2/2E \tag{1.2}$$

Table 1.1. *Some typical values of* $R_s, n, kR_s$ and $ka$

| Target | Proj. | $R_s$(fm) | $n$ | $kR_s$ | $ka$ |
|---|---|---|---|---|---|
| $^{16}$O | $\alpha$ | 6.2 | 0.9 | 12.0 | 1.0 |
| | $^{12}$C | 7.2 | 2.7 | 30.0 | 2.1 |
| $^{40}$Ca | $\alpha$ | 7.5 | 2.2 | 17.0 | 1.1 |
| | $^{12}$C | 8.6 | 6.7 | 49.0 | 2.8 |
| | $^{40}$Ca | 10.3 | 22.5 | 127.0 | 6.1 |
| $^{58}$Ni | $\alpha$ | 8.2 | 3.1 | 19.0 | 1.1 |
| | $^{12}$C | 9.2 | 9.4 | 56.0 | 3.0 |
| | $^{40}$Ca | 10.9 | 31.5 | 158.0 | 7.2 |
| $^{208}$Pb | $\alpha$ | 11.3 | 9.2 | 27.0 | 1.2 |
| | $^{12}$C | 12.3 | 27.7 | 86.0 | 3.5 |
| | $^{40}$Ca | 14.0 | 92.2 | 289.0 | 10.3 |

where $E$ is the asymptotic kinetic energy of relative motion

$$E = \tfrac{1}{2}\mu v^2 = E_{\text{lab}} A_2/(A_1 + A_2)$$

$A_1$ is the mass number of the projectile and $A_2$ is that of the target.

As indicated earlier semi-classical methods are useful when the de Broglie wave length is small or when the wave number $k$ of relative motion is large. Three dimensionless variables useful for measuring the magnitude of $k$ are

(a) Sommerfeld's parameter $n = ka_C$ which compares the de Broglie wave length with the Coulomb length parameter $a_C$,
(b) $kR_s$ which compares it with the strong absorption radius,
(c) $ka$ which compares it with a surface diffuseness parameter (typically $a \approx 0.5$ fm).

Table 1.1 gives some values of these parameters for a range of projectiles and targets. In each case the laboratory energy of the projectile is $E_{\text{lab}}/A_1 =$ 8 MeV ($v/c = 0.13$). The product $kR_s$ is large for every combination of target and projectile in table 1.1. Sommerfeld's parameter $n$ is of the order of unity when the target and projectile are both light. In all other cases $n \gg 1$. Even the product $ka > 1$ for every pair. These numbers show that it should be possible to use short wave length approximations in analysing and describing nucleus–nucleus collisions.

Application of semi-classical ideas to nuclear scattering problems began with Blair's work on elastic scattering (Blair, 1954) and inelastic scattering (Blair, 1959) of $\alpha$-particles. These papers are based on a diffractive picture of the reactions and identify Fresnel and Fraunhofer-type angular distributions for elastic scattering. A series of papers by Frahn and collaborators which was initiated by Frahn & Venter (1963) has had a major influence on the subject. This work emphasizes the importance of diffraction, especially of Fresnel diffraction for understanding heavy-ion elastic angular distributions. Frahn's overriding purpose was to obtain 'closed form' expressions for the elastic amplitude. His approach was not based on optical potentials but rather on a parametrization of the $S$-matrix elements $S_l$ in $l$-space. This is an attractive idea in view of the known fact that optical potentials are not unique as shown by the existence of different 'phase-equivalent' potentials. Frahn seems to have been the first to recognize the importance of the nearside–farside decomposition of the scattering amplitude. Frahn's work went through many stages of development. The best source for his work on elastic scattering is his last review paper (Frahn, 1984).

Semi-classical, as opposed to diffractive approaches to heavy-ion scattering are based on the WKB method (Jeffreys, 1923; Wentzel, 1926; Kramers, 1926; and Brillouin, 1926). Applications to scattering problems began with the classic work of Ford & Wheeler (1959). There is an excellent survey of semi-classical methods including application to scattering theory in a review by Berry & Mount (1972) and this paper had an important influence on subsequent developments. One of the most significant advances since that time was due to Knoll & Schaeffer (1976, 1977). Their work introduced the powerful and elegant methods of complex angular momentum, but did not lead to closed formulae like the ones obtained in Frahn's approach.

The Copenhagen group has made important contributions to the semi-classical theory of heavy-ion reactions. The book by Broglia & Winther (1981) contains an excellent survey of their results.

There is a conflict between the diffractive and semi-classical descriptions of heavy-ion scattering which is analogous to the conflict between Huygens' wave picture and Keller's (1955, 1957, 1958, 1962) ray picture of diffraction in optics. Indeed, the work of Knoll & Schaeffer is influenced strongly by Keller's ideas. The two approaches provide alternative descriptions which are compatible but not completely equivalent.

There are excellent surveys of the path integral method in the book by Feynman and Hibbs (1965) and in a recent book on *Techniques and Applications of Path Integration* by Schulman (1981). The advantage of this method for semi-classical analysis for systems with several separable degrees of freedom was already noticed by Gutzwiller (1967, 1969, 1970),

Pechukas (1969), Miller (1974, 1975) and others. The presentation in chapters 7 and 8 of this book is based on papers by Levit & Smilansky (1977*a*, *b*).

A general review of many aspects of heavy-ion reactions is contained in the book by Bass (1980). Other material is available in the reports of various conferences and schools.

The treatment of semi-classical methods in nucleus–nucleus collisions presented in this book is limited in its scope. The emphasis is on elastic scattering. The discussion of inelastic scattering in chapter 9 does not go into so much detail. Other important topics like polarization, deep inelastic scattering and the application of the methods of the Vlasov equation and semi-classical fluid dynamics have been omitted.

# 2

# Elastic scattering and the optical model

## 2.1 Coulomb scattering

In many cases the nuclei involved in a heavy-ion collision have large charges and Coulomb forces are very important. When the incident energy is sufficiently low the nuclei do not come close together during the collision and only the Coulomb force acts between them and, if classical theory applies, the relative motion follows a Rutherford orbit (fig. 2.1). The relations between the impact parameter $b$, the scattering angle $\theta$ and the distance of closest approach $d$ for a Rutherford orbit are

$$b = a_C \cot \tfrac{1}{2}\theta, \quad d = a_C + (a_C^2 + b^2)^{\frac{1}{2}} \tag{2.1}$$

where $a_C$, the Coulomb length parameter, is given by equation (1.2). The relations (2.1) can be written in terms of dimensionless parameters as

$$\lambda = n \cot \tfrac{1}{2}\theta, \quad kd = n + (n^2 + \lambda^2)^{\frac{1}{2}} \tag{2.2}$$

where $\lambda = kb$ is the relative angular momentum of the target and projectile measured in units of $\hbar$ and $n$ is Sommerfeld's parameter

$$n = ka_C = Z_1 Z_2 e^2/\hbar v \tag{2.3}$$

The Rutherford cross-section is

$$\sigma_R(\theta) = \tfrac{1}{4}a_C^2(\operatorname{cosec} \tfrac{1}{2}\theta)^4 \tag{2.4}$$

The quantum mechanical Coulomb scattering problem can be solved exactly (Messiah (1959) ch. 11, Landau & Lifshitz (1958) ch. 14). When the relative coordinate $\mathbf{r}$ is large the wave function $\psi$ describing the scattering is the sum of two parts

$$\psi = \psi_i + \psi_{sc}$$

The part $\psi_i$ represents the incident wave. It is a plane wave with a modification due to the long range character of the Coulomb potential

$$\psi_i(\mathbf{r}) \sim \exp [i(kz + n \ln k(r - z))]$$

if the incident wave is travelling in the positive $z$-direction. The scattered wave $\psi_{sc}$ has the following asymptotic form for large $r$

$$\psi_{sc}(\mathbf{r}) \sim (1/r) \exp [i(kr - n \ln (2kr))] f_C(\theta)$$

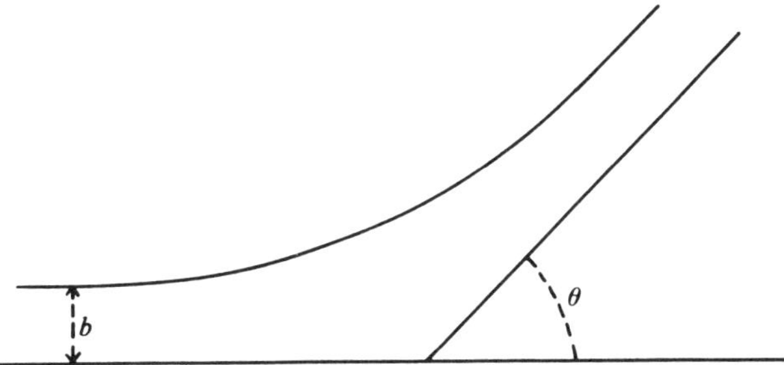


Fig. 2.1. A Rutherford orbit showing the scattering angle $\theta$ and the impact parameter $b$.

where the Coulomb scattering amplitude

$$f_C(\theta) = -\frac{n}{2k\sin^2\frac{1}{2}\theta}\exp\left[-in\ln(\sin^2\tfrac{1}{2}\theta) + 2i\sigma_0\right] \tag{2.5}$$

and $\sigma_0 = \arg\Gamma(1 + in)$ is the Coulomb phase shift. The quantal Formula for the scattering cross-section

$$\sigma_R(\theta) = |f_C(\theta)|^2$$

gives the same result (2.4) as in the classical case.

## 2.2 Optical potentials

At higher energies the nuclei in a heavy-ion collision come closer together and nuclear forces can influence the scattering. The nuclear interaction in an elastic collision is usually described by an optical potential

$$V_n(r) = U_n(r) - iW(r) \tag{2.6}$$

The real part $U_n(r)$ represents the nuclear forces acting between the two nuclei while the imaginary part describes absorption of flux from the elastic scattering channel due to various competing reactions.

If $\psi(\mathbf{r}, t)$ is a wave function describing the relative motion of the two nuclei under the influence of the optical potential then

$$\rho(\mathbf{r}, t) = |\psi(\mathbf{r}, t)|^2$$

is the probability density for finding them in their ground states with relative coordinate $\mathbf{r}$ and

$$\mathbf{j}(\mathbf{r}, t) = (\hbar/2mi)(\psi^*(\mathbf{r})\nabla\psi(\mathbf{r}) - \psi(\mathbf{r})\nabla\psi^*(\mathbf{r}))$$

is a probability current. The wave- function $\psi(\mathbf{r}, t)$ satisfies a time-dependent Schrödinger equation, and it is easy to show that

$$\partial\rho/\partial t + \operatorname{div}\mathbf{j} = -(2W(r)/\hbar)\rho(\mathbf{r}) \tag{2.7}$$

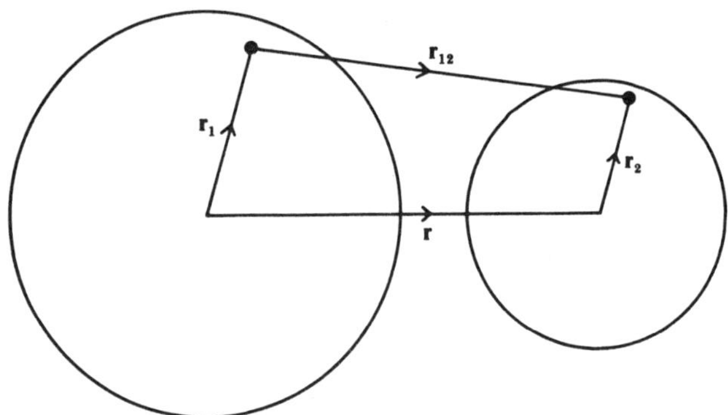


Fig. 2.2. Coordinates for the double-folded potential, equation (2.8).

The right-hand side of the equation (2.7) represents loss of probability from the elastic channel into other channels, and $2W(r)/\hbar$ is the transition probability per unit time for a transition out of the elastic channel when the relative coordinate of the two nuclei is $\mathbf{r}$. It is clear from this physical interpretation that $W(r)$ should be a positive quantity. This is the reason for writing a negative sign in equation (2.6).

A very simple model for the real part of the optical potential is constructed by assuming that the total potential energy of the interaction of two nuclei is obtained by summing the interactions between pairs of nucleons. This double-folding model has been reviewed by Satchler & Love (1979) and is used extensively to fit heavy-ion elastic scattering data. The double-folded potential may be written as

$$U_{\text{n}}(r) = \int d\mathbf{r}_1 \int d\mathbf{r}_2 \rho_1(\mathbf{r}_1)\rho_2(\mathbf{r}_2)v(\mathbf{r}_{12}) \tag{2.8}$$

Here $\rho_i$ is the density distribution of nucleons in the ground state of the $i$th nucleus while the coordinates are defined in fig. 2.2. In equation (2.8) $v(\mathbf{r}_{12})$ is an effective interaction potential between a nucleon in the first nucleus and one in the second. Their relative position vector is $\mathbf{r}_{12}$. Various simple choices of this effective interaction have been considered by Satchler & Love (1979). When analysing experimental data the folding potential (2.8) is usually multiplied by a factor $N$ which is varied and chosen to give a best fit to the data. Love & Satchler have studied many examples and find that $N \approx 1.11 \pm 0.13$ when $v$ is chosen to be the M3Y effective interaction developed by Bertsch *et al.* (1977).

The total nucleus–nucleus interaction potential is the sum of a nuclear and a Coulomb part

$$V(r) = V_{\text{n}}(r) + V_{\text{C}}(r)$$

Table 2.1. *Parameters for the E18 and GK Potentials*

| | $V_0$ | $r_v$ | $a_v$ | $W_0$ | $r_w$ | $a_w$ | $r_C$ |
|---|---|---|---|---|---|---|---|
| E18 | 10.0 | 1.35 | 0.615 | 23.4 | 1.23 | 0.552 | 1.0 |
| GK | 75.21 | 1.23 | 0.493 | 8.5 | 1.21 | 0.184 | 1.42 |

$V_0$ and $W_0$ in MeV, $r_v$, $r_w$, $r_C$, $a_v$ and $a_w$ in fm.

The best way to calculate the Coulomb part $V_C$ is to use a folding formula like (2.8) with $v$ replaced by the Coulomb repulsion between point charges and $\rho_1$ and $\rho_2$ replaced by the charge densities of the two nuclei. If the nuclei are spherical the general form of $V_C$ is

$$V_C(r) = K(r)Z_1Z_2e^2/r \tag{2.9}$$

where $K(r) = 1$ when their charge distributions do not overlap. In principle $K(r)$ can be calculated from the folding model, but de Vries & Clover (1975) have shown that the scattering is not sensitive to the detailed shape of $K(r)$. In many applications the following form is used

$$K(r) = 1; \qquad r > R_C$$
$$K(r) = (\tfrac{3}{2}(r/R_C) - \tfrac{1}{2}(r/R_C)^3); \qquad r < R_C$$

The Coulomb radius $R_C = r_C(A_1^{\frac{1}{3}} + A_2^{\frac{1}{3}})$ where $r_C$ is chosen in the range 1 fm to 1.3 fm.

The expression (2.8) for the folding potential involves a six-dimensional integral, but Satchler & Love (1979) point out that it is relatively simple to evaluate. They recommend a Fourier transform method which is very fast to implement on a computer. However, because of the extra calculation required for the folding potential, many elastic scattering studies use simple parametrized forms for $U_n$. The most popular choice is a Woods–Saxon potential

$$U_n(r) = -V_0/[1 + \exp((r-R)/a)] \tag{2.10}$$

which is specified by a strength $V_0$, a range $R$ and a surface diffuseness $a$. Often, the imaginary part $W(r)$ of the optical potential is also parametrized by a Woods–Saxon form (2.10), but the radius and surface diffuseness parameters are not necessarily the same as for the real part. Table 2.1 shows two examples of sets of Woods–Saxon parameters which have been used to fit experimental elastic scattering cross-sections of $^{16}O$ by $^{28}Si$. The E18 potential gives a good fit to forward angle scattering over a range of energies (Cramer *et al.*, 1976), while the GK potential was fitted to the scattering at $E_{lab} = 55$ MeV over a wide range of angles (Takemasa & Tamura, 1978).

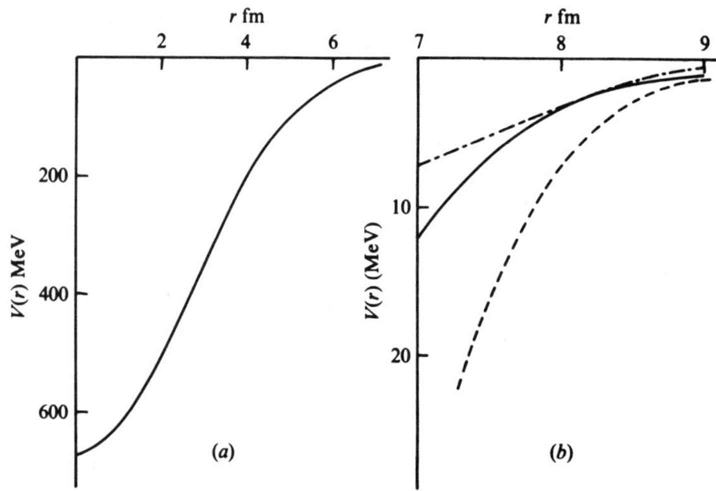


Fig. 2.3. (*a*) The folded potential for $^{16}O + ^{28}Si$ calculated with the M3Y interaction, (*b*) comparison with the E18 and GK potentials in the surface.

Fig. 2.3 shows a sketch of the folded potential for $^{16}O + ^{28}Si$ and a comparison of the folded potential with the phenomenological potentials E18 and GK for separations $r$ near the strong absorption radius $R_s$. Several points should be noted:

(a) The potentials are very different for small $r$. In particular, the folded potential is much deeper than either of the phenomenological potentials.

(b) All three potentials have similar strenghts for $r$ near the strong absorption radius. The three potentials have comparable magnitudes for $r$ between 8 fm and 10 fm.

Many studies have shown that optical model parameters are not well determined by experimental data. Ball *et al.* (1975) studied several examples in detail and concluded that many quite different sets of parameters give equally good fits to the experimental results. In particular, changing $V_0$, $W_0$, $R_v$, and $R_w$ keeping

$$V_0 \exp(R_v/a_v) = \text{const.}, \quad W_0 \exp(R_w/a_w) = \text{const.} \tag{2.11}$$

often produces very little change in the cross-section. This is an example of an ambiguity noted by Igo in 1953. The conditions (2.11) applied to a Woods–Saxon potential imply that the interaction is unchanged for large values of $r$, even though there can be big changes for small $r$. The existence of an Igo ambiguity indicates that the scattering is sensitive to the form of $V_n(r)$ only for large $r$ values. Ball *et al.* (1975) have shown that for scattering of $^{16}O$ by $^{208}Pb$ the cross-section is sensitive to the optical potential mainly for values of $r$ near the strong absorption radius (1.1).

In the case of $^{16}O$ on $^{28}Si$ the scattering is somewhat sensitive to the detailed shape of the optical potential. The examples E18 and GK show that quite different potentials can be used to fit data for the same target and projectile combination. Here, however, the data fitted is not the same. E18 fits forward angle data for projectile energies varying from 20 MeV to 150 MeV. On the other hand, GK fits the elastic scattering angular distribution at forward and backward angles, but only at one energy. More recent studies (Kobos & Satchler, 1984) show that the $^{16}O$ on $^{28}Si$ scattering is sensitive to some of the details of the optical potential for $r$ down to 6 fm.

In other cases where the scattering is sensitive to the shape of the optical potential phenomenological forms which are different from the Woods–Saxon have been proposed. For example, Michel & Vanderpoorten (1977) have used a shape which is a power of a Woods–Saxon

$$U_n(r) = -V_0/[1 + \exp((r - R)/va)]^v \tag{2.12}$$

to fit some $(\alpha, {}^{40}Ca)$ angular distributions. If $v = 2$ the radial dependence is very similar to that of potentials calculated from the folding model. Another parametrization suitable for systems like $(\alpha, {}^{16}O)$ has been proposed by Buck & Pilt (1977)

$$U_n(r) = -\frac{V_0(1 + \cosh(R/a))}{\cosh(r/a) + \cosh(R/a)} \tag{2.13}$$

When $R \gg a$ the potential (2.13) is very similar to a Woods–Saxon, but by suitably choosing $R$ and $a$ potential shapes can be obtained which are remarkably similar to folding potentials.

For heavier systems the elastic scattering is sensitive to the optical potential mainly for $r$ near the strong absorption radius. There is a simple systematic variation of the potential from one pair of nuclei to another, which can be derived by an argument due to Blocki *et al.* (1977). The idea is this: when the nuclear density distributions are not overlapping the force between the two nuclei is due mainly to the interaction between some nucleons in the surface of one nucleus with some nearby nucleons in the surface of the other. Starting from this idea Blocki *et al.* show that the interaction energy of two nuclei with radii $R_1$ and $R_2$ whose centres are separated by a distance $r = R_1 + R_2 + S$ is given approximately by

$$U_n(r) = 2\pi \frac{R_1 R_2}{R_1 + R_2} \int_S^\infty e(t)\, dt \tag{2.14}$$

In this equation $e(S)$ is a universal function which is the same for all pairs of nuclei. It is the interaction energy per unit area of two plane semi-infinite sheets of nuclear matter with parallel surfaces separated by a distance $S$. The form (2.14) is called a proximity potential. The approximations made in deriving it are accurate if $S < R_1, R_2$, i.e. if the separation between the surfaces of the nuclei is small compared with their radii.

The scattering by a strongly absorbing potential is sensitive to the form of the optical potential only for large separations. Then a Woods–Saxon potential like (2.10) can often be replaced by its exponential tail

$$U_n(r) \simeq -V_0 \exp((R - r)/a) \tag{2.15}$$

The real potentials obtained from a large number of analyses of elastic heavy-ion scattering have been summarized by Christensen & Winther (1976) in terms of such an exponential potential

$$U_n(r) = -50[R_1 R_2/(R_1 + R_2)] \exp((R_1 + R_2 - r)/a)\,\text{MeV} \tag{2.16}$$
$$R_i = 1.223 A_i^{\frac{1}{3}} - 0.978 A_i^{-\frac{1}{3}}\text{fm}$$

where $a = 0{\cdot}63$ fm. This potential, which has the proximity form (2.14), gives a good account of the average behaviour of potentials which fit observed scattering, although minor fluctuations in the values will be found in individual cases.

## 2.3 Scattering amplitude

The differential scattering cross-section

$$\sigma(\theta) = |f(\theta)|^2$$

is calculated from the scattering amplitude $f(\theta)$. This in turn is given by the partial wave formula

$$f(\theta) = 1/(2ik) \sum_l (2l + 1) P_l(\cos\theta)(S_l - 1)$$

where the $S_l(E)$ are the elastic partial wave amplitudes. Because of the strong Coulomb interaction between heavy-ions it is convenient to separate the Coulomb contribution by writing

$$S_l(E) = S_{nl} \exp(2i\sigma_l) \tag{2.17}$$

The Coulomb phase shifts are given by

$$\sigma_l = \arg \Gamma(l + 1 + in) \tag{2.18}$$

when $n$ is the Sommerfeld parameter (2.3). When there is no nuclear interaction $S_{nl} = 1$. The scattering amplitude $f(\theta)$ can be separated into a Coulomb and a nuclear part by writing

$$f(\theta) = f_C(\theta) + 1/(2ik)\sum(2l+1)P_l(\cos\theta)\exp(2i\sigma_l)(S_{nl}-1) \qquad (2.19)$$

where $f_C(\theta)$ is given by equation (2.5). The nuclear partial wave amplitude is related to the nuclear phase shift $\delta_{nl}$ by

$$S_{nl} = \exp(2i\delta_{nl}) \qquad (2.20)$$

This, in turn, is determined by the asymptotic form for larger $r$ of the radial wave function $F_l(r)$ of the relative motion of the target and projectile

$$F_l(r) \sim \sin(kr - \tfrac{1}{2}l\pi + n\ln 2kr + \sigma_l + \delta_{nl}) \qquad (2.21)$$

The usual optical model procedure for obtaining the phase shifts is to solve the radial Schrödinger equation for $F_l(r)$ numerically

$$\frac{\hbar^2}{2\mu}\left(-\frac{d^2}{dr^2} + \frac{l(l+1)}{r^2}\right)F_l(r) + V(r)F_l(r) = EF_l(r) \qquad (2.22)$$

to match the numerical solution on to a Coulomb wave function at some point $r_0$ outside the range of the nuclear optical potential and then to extract the phase shift by comparison with the asymptotic form (2.21). For a real potential $V(r)$ the nuclear phase $\delta_{nl}$ is real and $|S_{nl}| = 1$. If the optical potential has an absorptive imaginary part as discussed in section (2.2) then $\operatorname{Im}\delta_{nl} \geqslant 0$ and $|S_{nl}| \leqslant 1$.

## 2.4 Fresnel-type scattering

Heavy-ion elastic scattering angular distributions have a very characteristic shape when the Sommerfeld parameter $n \gg 1$. Here we show some results for scattering of $^{12}C$ by $^{208}Pb$ for $E_{lab} = 96$ MeV. In this example $n = 19.1$. The experimental data is given by Ball *et al.* (1975). The authors compare their results with angular distributions calculated from various choices of optical potentials. Fig. 2.4 shows the ratio of the elastic cross-section $\sigma(\theta)$ to the Rutherford cross-section $\sigma_R(\theta)$ plotted against the centre-of-mass scattering angle $\theta$. The curve is the result of an optical model calculation. The optical potential is a Woods–Saxon

$$V_n(r) = -(V_0 + iW_0)/[1 + \exp((r-R)/a)]$$

with strength parameter $V_0 = 40$ MeV and $W_0 = 25$ MeV. The radius parameter is $R = 8.21$ fm and the surface diffuseness is $a = 0.56$ fm for both the real and the imaginary part.

Fig. 2.5 is a sketch of the magnitude $|S_l| = \exp(-2\operatorname{Im}\delta_{nl})$ of the partial

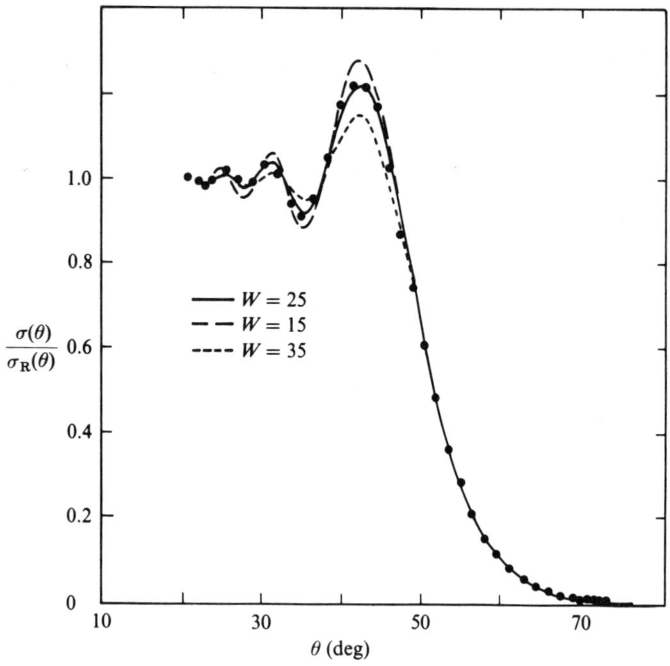


Fig. 2.4. Angular distribution for $^{12}C$ on $^{208}Pb$ at $E_{lab} = 96$ MeV. Comparison between theory and data for three values of the imaginary strength $W$ (in MeV). The other parameters are given in the text (from Ball *et al.*, 1975).

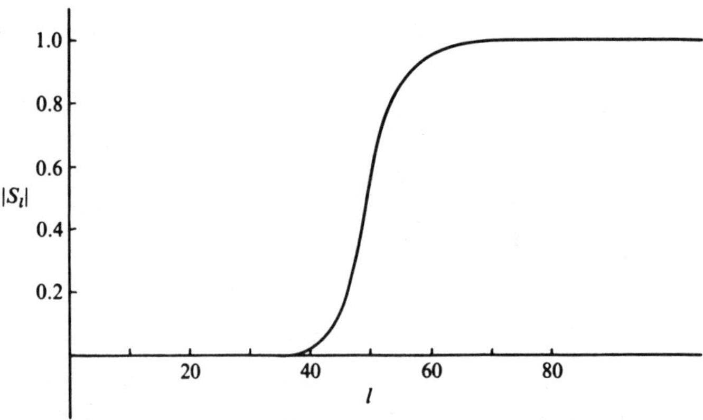


Fig. 2.5. Magnitude of the partial wave amplitude $|S_l|$ for $^{12}C$ on $^{208}Pb$ at $E_{lab} = 96$ MeV. The parameters are the same as for fig. 2.4 with $W = 25$ MeV.

wave scattering amplitude for the same example. When the angular momentum $l$ is large enough the projectile does not come close enough to the target during the collision for the nuclear forces to act, the nuclear phase shift $\delta_{nl} \approx 0$ and $|S_l| \approx 1$. This is the situation for $l \gtrsim 50$ in fig. 2.5. For small $l$ the projectile interacts with the target very strongly and $|S_{nl}| \approx 0$. Fig. 2.5 shows that partial waves with $l < 50$ are absorbed very strongly.

A very simple model proposed by Blair (1954) reproduces some of the main features of the angular distribution in Fig. 2.4. This is the 'sharp-cut-off model' and it makes the simplest and most extreme assumption about strong absorption in $l$-space

$$\begin{aligned} S_l &= 0; & l &\leqslant l_g \\ S_l &= \exp(2i\sigma_l); & l &> l_g \end{aligned} \tag{2.23}$$

where $\sigma_l$ are the Coulomb phase shifts. The model assumes that all partial waves up to a grazing angular momentum are completely absorbed, while all higher ones undergo Rutherford scattering. When $l_g \gg 1$ and $n \gg 1$ the partial wave series with the strong absorption amplitudes (2.23) can be summed approximately (Frahn, 1966) to give

$$f(\theta)/f_C(\theta) \simeq \tfrac{1}{2}[1 - (C(w) + S(w)) - i(C(w) - S(w))] \tag{2.24}$$

where $C(w)$ and $S(w)$ are Fresnel integrals and

$$w = [l_g/(\pi \sin\theta_g)]^{\frac{1}{2}}(\theta - \theta_g) \tag{2.25}$$

$$l_g = n \cot(\tfrac{1}{2}\theta_g) \tag{2.26}$$

The amplitude (2.24) gives a ratio of the cross-section to the Rutherford cross-section

$$\sigma(\theta)/\sigma_R(\theta) = \tfrac{1}{2}[(\tfrac{1}{2} - C(w))^2 + (\tfrac{1}{2} - S(w))^2] \tag{2.27}$$

The right-hand side of equation (2.27) is formally identical to the Fresnel diffraction pattern of classical optics. Fig. 2.6 shows the comparison of the predictions of the simple Fresnel formula (2.27) with experimental data for $^{16}O + {}^{208}Pb$ elastic scattering.

Two angular regions can be distinguished in the Fresnel cross-section. When $\theta < \theta_g$ the ratio $\sigma/\sigma_R$ oscillates about a mean value of unity. The oscillations are due to interference between a Coulomb scattering amplitude (2.5) and a diffracted amplitude from the sharp cut-off in $S_l$ at $l = l_g$. The experimental data and the optical model calculations show similar oscillations, but these are not in phase with the predictions of the Fresnel formula (2.27). This is because the derivation of (2.27) contains some rather crude approximations. A more refined formula described in chapter 5 gives

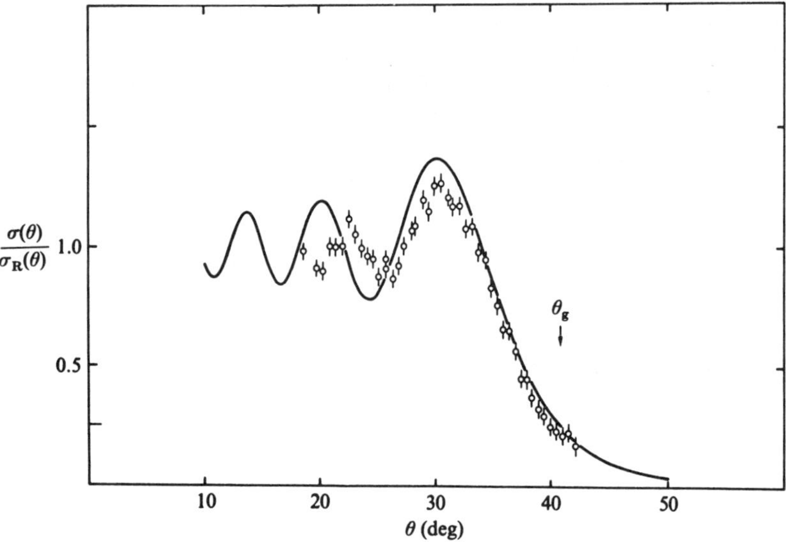


Fig. 2.6. Fresnel diffraction effects in elastic scattering of $^{16}O$ by $^{208}Pb$ at $E_{lab} = 166$ MeV. The curve is calculated from equation (2.27) with grazing angular momentum $l_g = 86$. The Sommerfeld parameter is $n = 32.05$ and $\theta_g = 41°$ (from Frahn, 1966).

a much more accurate representation of the cross-section in this angular region.

When $\theta > \theta_g$ the Coulomb amplitude is absent due to the strong absorption of waves with $l < l_g$, and only the diffracted wave is present. The Fresnel formula (2.27) predicts that $\sigma/\sigma_R = \frac{1}{4}$ at the classical grazing angle $\theta_g$, which is related to the cut-off angular momentum $l_g$ by equation (2.26). The quarter-point property was first found in $\alpha$-particle scattering and has been used in the 'quarter point recipe' of Blair (1954) to determine the cut-off angular momentum directly from the experimental data. The quarter-point angle for ($^{12}C$, $^{208}Pb$)-scattering at 96 MeV is $\theta_g = 56°$, and the Rutherford formula (2.32) gives a grazing angular momentum $l_g \simeq 50$.

There is an alternative to the Fresnel diffractive interpretation of the angular distribution in fig. 2.4. Some authors argue that this kind of angular distribution is produced by refractive effects of the real nuclear potential. The refractive interpretation (Broglia *et al.*, 1974a; Knoll & Schaeffer, 1976) says that Coulomb orbits, which would have been scattered into the shadow region, are pulled forward by the attractive real nuclear potential. A concentration of such rays results in a rainbow peak which is identified with the largest peak in fig. 2.4 and has been called the Coulomb rainbow. We discuss the refractive rainbow interpretation in section 4.4 and the Fresnel interpretation in chapter 5.

For convenience we will often refer to this kind of scattering pattern as Fresnel type.

## 2.5 Fraunhofer scattering

The Sommerfeld parameter gives a rough guide to the importance of the Coulomb interaction between the nuclei in a heavy-ion scattering experiment. If $n \gg 1$ the Coulomb force is very important and the elastic cross-section has the Fresnel diffraction structure discussed in section 2.4. When $n \lesssim 1$ the Coulomb force is less effective and can be neglected in the simplest qualitative theory of the scattering process. Fig. 2.7 shows some experimental data for elastic scattering of $^{16}O$ by $^{12}C$ at $E_{\text{lab}} = 168$ MeV (Hiebert & Garvey, 1964). In this example, the Sommerfeld parameter $n = 2.3$.

A very simple theory is obtained by neglecting the Coulomb interaction in the strong absorption model so that the partial wave amplitude $S_l$ is given by

$$\left.\begin{aligned} S_l &= 0; \quad l < l_g \\ S_l &= 1; \quad l \geqslant l_g \end{aligned}\right\} \tag{2.28}$$

When $l_g \gg 1$ and $\theta$ is not too large the scattering amplitude corresponding to these partial wave amplitudes can be evaluated approximately with the result (Blair, 1959)

$$f(\theta) \simeq i(l_g + \tfrac{1}{2})J_1((l_g + \tfrac{1}{2})\theta)/(k\theta) \tag{2.29}$$

where $J_1$ is a Bessel function. The cross-section is

$$\sigma(\theta) \simeq [(l_g + \tfrac{1}{2})J_1((l_g + \tfrac{1}{2})\theta)/(k\theta)]^2 \tag{2.30}$$

and is the Fraunhofer diffraction pattern produced by an absorbing disk with radius $R = (l_g + \frac{1}{2})/k$. The angular spacing between successive maxima is

$$\Delta\theta \simeq \pi/(l_g + \tfrac{1}{2}) \tag{2.31}$$

The dashed curve in fig. 2.7 shows the prediction of equation (2.30). The experimental data also shows a clear diffraction structure, but the Fraunhofer formula (2.30) does not predict the details correctly. The diffraction oscillations in the experiment are less pronounced than the predictions of the sharp cut-off model. More detailed calculations in chapter 5 show

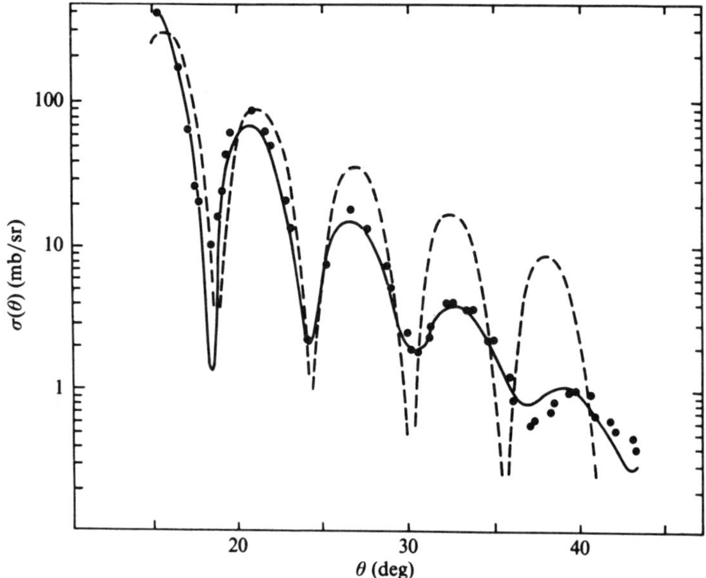


Fig. 2.7. Elastic scattering of $^{16}O$ on $^{12}C$ at $E_{lab} = 168$ MeV. The points show experimental data and the solid line is calculated from an optical model. The dashed line is from the plane wave Born approximation which resembles the diffraction formula (2.30) (from Hiebert & Garvey, 1964).

that modifications of the sharp cut-off model and the Coulomb effects combine to damp out the diffraction oscillations. Nevertheless, the prediction (2.31) of the spacing $\Delta\theta$ between successive maxima of the diffraction pattern is not changed when the theory is improved and it can be used to extract a value for the grazing angular momentum directly from the experimental data. The spacing of the oscillations in fig. 2.7 gives $l_g \simeq 30$. This corresponds to an interaction radius $R_s = (l_g + \frac{1}{2})/k = 6.2$ fm.

## 2.6 The strong absorption radius

The terms 'grazing angular momentum' and 'strong absorption radius' have been used in the introduction and in sections 2.4 and 2.5. They have a well-defined meaning in the sharp cut-off model (2.23) and (2.28). There is no unique definition when the nucleus–nucleus interaction is represented by an optical potential, although prescriptions in use give values which differ usually only by a few percent.

There are two empirical prescriptions, one based on Fresnel diffraction ideas and the other on the Fraunhofer diffraction picture. In section 3.6

it is shown that the orbital angular momentum quantum number $l$ is associated with a semi-classical angular momentum $\lambda = l + \frac{1}{2}$. The scattering angle $\theta$ and distance of closest approach $D$ for a Rutherford orbit are related by

$$\lambda = l + \tfrac{1}{2} = n \cot \tfrac{1}{2}\theta \tag{2.32}$$

$$\lambda^2 = (l + \tfrac{1}{2})^2 = kD(kD - 2n) \tag{2.33}$$

If an angular distribution is of Fresnel type as discussed in section 2.4 then Blair's quarter-point angle $\theta_{\frac{1}{4}}$ is the angle at which the measured cross-section has fallen to $\frac{1}{4}$ of the Rutherford value. The strong absorption radius may be defined as the distance of closest approach $D_{\frac{1}{4}}$ in a Rutherford orbit with scattering angle $\theta_{\frac{1}{4}}$. The corresponding grazing angular momentum $\lambda_{\frac{1}{4}}$ is related to $\theta_{\frac{1}{4}}$ by (2.32). A simple theory of a Fraunhofer-type angular distribution (2.30) gives a cross-section proportional to $(J_1(x)/x)^2$ with $x = \lambda\theta$. The spacing between the peaks of the oscillations in the cross-section is $\Delta x \simeq \pi$ and the minima correspond to zeros of the Bessel function $J_1(x)$, $x_1 = 3.83$, $x_2 = 7.01$, $x_3 = 10.17$. A grazing angular momentum $\lambda_{\mathrm{F}} = l_{\mathrm{F}} + \frac{1}{2}$ can be extracted either from the peak spacing $\Delta\theta$

$$\lambda_{\mathrm{F}} = l_{\mathrm{F}} + \tfrac{1}{2} \simeq \pi/\Delta\theta \tag{2.34}$$

or, more accurately, from the minima in the cross-section. The strong absorption radius can be defined as $D_{\mathrm{F}}$ related to $\lambda_{\mathrm{F}}$ by (2.33).

The optical model or parametrized $S$-matrix can be used to analyse elastic scattering data of either the Fresnel or Fraunhofer type. However, there is no complete agreement on what is the most useful quantity to designate as the strong absorption radius. A popular prescription is to define a grazing angular momentum $\bar{l}_{1/2}$ by the property that the probability of reflection is 50%.

$$|S_{\bar{l}_{\frac{1}{2}}}|^2 = \tfrac{1}{2}, \quad |S_{\bar{l}_{\frac{1}{2}}}| = 1/\sqrt{2} \tag{2.35}$$

The strong absorption radius is obtained from $\bar{l}_{1/2}$ through the classical relation (2.33). Another popular choice is (Frahn & Rehm, 1978) to take $l_{1/2}$ as the $l$-value for which $|S_l| = \frac{1}{2}$. The latter prescription generally results in a slightly smaller stong absorption radius. Authors who favour a refractive interpretation rather than one based on Fresnel diffraction emphasize the distance of closest approach of the 'rainbow orbit' (see section 4.4).

Strong absorption radii obtained by any of the above methods are given approximately by

$$R_{\mathrm{s}} \simeq 1.5(A_1^{\frac{1}{3}} + A_2^{\frac{1}{3}})\,\mathrm{fm} \tag{2.36}$$

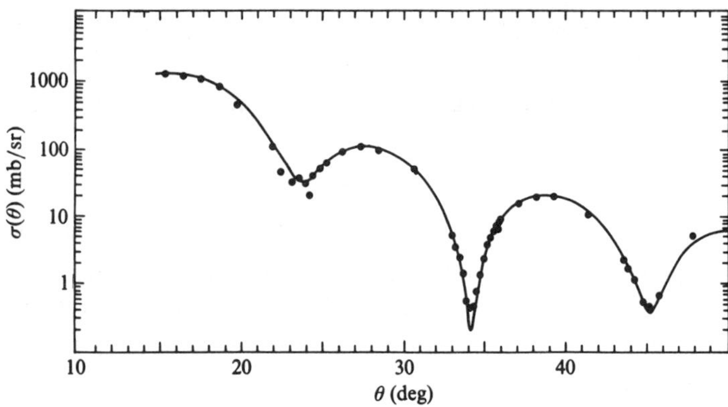


Fig. 2.8. Differential cross-section for scattering of 41.76 MeV α-particles from $^{44}$Ca. The points show experimental data and the curve is an optical model fit (from Fernandez & Blair, 1969).

Table 2.2. *Strong absorption radii in fm for alpha-scattering from several nuclei (from Fernandez & Blair, 1969)*

| Target | $E_{lab}$ | $D_F$ | $\bar{D}_{\frac{1}{2}}$ | $D_{\frac{1}{2}}$ | $D_0$ |
|---|---|---|---|---|---|
| $^{40}$Ca | 41.78 | 7.281 | 7.461 | 7.360 | 7.416 |
| $^{42}$Ca | 41.76 | 7.329 | 7.519 | 7.385 | 7.464 |
| $^{44}$Ca | 41.76 | 7.383 | 7.590 | 7.429 | 7.493 |
| $^{48}$Ca | 41.77 | 7.404 | 7.662 | 7.498 | 7.529 |
| $^{54}$Fe | 42.02 | 7.635 | 7.833 | 7.693 | 7.764 |
| $^{58}$Ni | 42.00 | 7.848 | 8.196 | 7.897 | 7.974 |

Notation is defined in the text.

Another approximate formula suggested by Schröder & Huizenga (1977) is

$$R_s \simeq 1.1(A_1^{\frac{1}{3}} + A_2^{\frac{1}{3}}) + 3\,\text{fm} \tag{2.37}$$

These values of $R_s$ are large. The half-density points of the two nuclei are separated by roughly 3 fm. This corresponds to a very small overlap of the two densities.

Fig. 2.8 shows an experimental angular distribution for elastic scattering of 41.76 MeV particles from $^{44}$Ca together with the best-fit cross-sections of an optical model and a parametrization of the scattering amplitudes (Fernandez & Blair, 1969). The cross-section is of Fraunhofer type and

Fernandez & Blair have extracted a Fraunhofer grazing angular momentum $l_F$ from the position of the deep minimum in the cross-section at 34.14° corresponding to the zero at $x_3 = 10.17$ of the Bessel function. They have also obtained $\bar{l}_{1/2}$ and $l_{1/2}$ from the optical model partial wave amplitudes and an orbiting angular momentum $\lambda_0$ from the condition that the top of the barrier in the effective radial potential which is the sum of nuclear, Coulomb and centrifugal parts coincides with the centre-of-mass energy. They extract strong absorption radii $D_F$, $\bar{D}_{1/2}$, $D_{1/2}$ and $D_0$ from these angular momenta using equation (2.33). Table 2.2 shows these radii for the case illustrated in fig. 2.8 and for other cases studied by Fernandez & Blair. The numbers give an indication of the variation in strong absorption radii obtained from different prescriptions.

In the following chapters we use the terms 'grazing angular momentum $l_g$', 'grazing angle $\theta_g$' and 'strong absorption radius' in a general sense to stand for any one of these prescriptions. They are always related by the classical Rutherford equations (2.32) and (2.33). If we need to be more precise it will be specified in the text.

# 3

# The semi-classical phase-shift

## 3.1 The simple Wentzel, Kramers & Brillouin approximation

In the optical model approach to heavy-ion elastic scattering described in chapter 2 cross-sections are calculated by using numerical methods to solve the radial Schrödinger equation and sum the partial wave series. The numerical calculations present no problems, but it can be difficult to interpret the results. The relation between features of angular distributions and excitation functions and the optical potential can only be understood by making a systematic study of the ways in which variations of the potential influence the calculated cross-sections (Satchler, 1983*a, b*; Satchler & Love, 1979; Ball *et al.*, 1975). The semi-classical approach uses analytical methods. The scattering amplitude can be written as a superposition of a number of components, each of which has a clear physical origin and it is possible to obtain a better qualitative understanding of relations between nuclear potentials, partial wave amplitudes and measured cross-sections.

The wave length of relative motion of two nuclei in a heavy-ion reaction is small and semi-classical methods exploit the smallness of this quantity to obtain simple approximate formulae for scattering phase shifts and partial wave amplitudes.

If the de Broglie wave length of a particle is small compared with some characteristic dimension of a system then its properties are close to being classical and approximate semi-classical solutions of the Schrödinger equation can be found. The method is to use the WKB approximation which is discussed in many standard texts on quantum mechanics (Landau & Lifshitz, 1958; Messiah, 1959). We begin by collecting together the main results used in this book.

To obtain the WKB solutions we write the Schrödinger equation for a particle with mass $m$ and total energy $E$ moving in a potential $V(x)$ as

$$\frac{d^2\psi}{dx^2} + k^2(x)\psi = 0 \tag{3.1}$$

where $k^2(x) = (2m/\hbar^2)(E - V(x))$. Hence, $k(x)$ is the local wave number and

$p(x) = \hbar k(x)$ is the local momentum of the particle. Making the substitution

$$\psi(x) = \exp\left( i \int \eta(x) dx \right)$$

so that $\psi' = i\eta\psi$ and $\psi'' = i(\eta'\psi + \eta\psi')$ equation (3.1) reduces to the Ricatti equation

$$\eta^2 = k^2(x) + i d\eta/dx \tag{3.2}$$

The WKB approximation is found by assuming that $\eta(x)$ is a slowly varying function of $x$ in the sense that $|\eta'| \ll |\eta|^2$ and solving equation (3.2) by iteration. Iterating twice gives

$$\eta(x) \simeq \pm k(x) + \tfrac{1}{2} i k'/k \tag{3.3}$$

or

$$\psi(x) = C_1 k^{-\frac{1}{2}} \exp\left( i \int k(x) dx \right) + C_2 k^{-\frac{1}{2}} \exp\left( -i \int k(x) dx \right) \tag{3.4}$$

This is the WKB wave function in a classically allowed region where $E > V(x)$ and $k(x)$ is real. In a classically forbidden region ($E < V(x)$) the wave number $k(x)$ is pure imaginary and the WKB solution analogous to (3.4)

$$\psi(x) = \bar{C}_1 \gamma^{-\frac{1}{2}} \exp\left( -\int \gamma(x) dx \right) + \bar{C}_2 \gamma^{-\frac{1}{2}} \exp\left( \int \gamma(x) dx \right) \tag{3.5}$$

is obtained by putting $k(x) = i\gamma(x)$ or

$$\gamma^2(x) = (2m/\hbar^2)(V(x) - E)$$

The approximation (3.3) is valid only if

$$|k'(x)| \ll |k(x)|^2 \quad \text{or} \quad |d\lambda/dx| \ll 1 \tag{3.6}$$

where $\lambda(x) = 1/k(x)$. Qualitatively this means that the wave length of a particle should change only slightly over distances of the order of one wave length.

The first term in equation (3.4) represents a wave travelling in the positive $x$-direction while the second term is a wave propagating in the opposite direction. The probability current density associated with the wave can be obtained from the formula

$$j(x) = \frac{\hbar}{2mi}(\psi^* d\psi/dx - \psi d\psi^*/dx) \tag{3.7}$$

When calculating the derivatives in (3.7) the condition (3.6) must be remembered. Thus

$$d\psi/dx = ik^{\frac{1}{2}}\left(C_1 \exp\left(i\int k(x)dx\right) - C_2 \exp\left(-i\int k(x)dx\right)\right) \tag{3.8}$$

Terms of order $k'/k^2$ compared with the leading term have been discarded. This must be done in order to obtain a consistent approximation. Substituting in (3.7) gives

$$j = (\hbar/m)(|C_1|^2 - |C_2|^2) \tag{3.9}$$

The two terms in the WKB wave function (3.5) in a classically inaccessible region are exponentially decreasing and increasing respectively. The probability current is

$$j = (\hbar/mi)(\bar{C}_1^*\bar{C}_2 - \bar{C}_1\bar{C}_2^*) \tag{3.10}$$

and can be non-zero if $\bar{C}_1$ and $\bar{C}_2$ are complex.

## 3.2 Asymptotic approximations

Suppose the potential $V(x)$ in (3.1) is slowly varying and $l$ is a characteristic distance over which it changes appreciably. Then $dk/dx \simeq k/l$. For the WKB method to hold the condition (3.6) is

$$|k/l| \ll k^2 \quad \text{or} \quad |kl| \gg 1 \tag{3.11}$$

It can be shown that the WKB wave functions are the first order approximation in an expansion in powers of $1/kl$. As $1/kl = \hbar/pl$ where $p$ is the classical momentum the corrections can also be considered as an expansion in powers of $\hbar$. It turns out that the series expansion so obtained is of the asymptotic rather than the ordinary type, and we now discuss this (Migdal, 1977).

Let $S_n(z)$ be a partial sum to $n$ terms of a series

$$S_n(z) = a_0 + a_1/z + a_2/z^2 + \cdots + a_n/z^n$$

Suppose that for fixed $n$ the series $S_n(z)$ gives an approximation to some function $f(z)$ which becomes better as $z$ becomes larger in the sense that

$$z^n(S_n(z) - f(z)) \to 0 \qquad \text{as } z \to \infty$$

then $S_n(z)$ is called an asymptotic expansion of the function $f(z)$ and one writes

$$f(z) \sim a_0 + a_1/z + a_2/z^2 + \cdots \quad \text{as } z \to \infty \tag{3.12}$$

The series (3.12) may diverge for given $z$ as $n \to \infty$, but even if it does it still gives some information about $f(z)$ because for any $n$ it gives a good description of $f(z)$ for large $z$. An ordinary convergent series tends to $f(z)$ when $n \to \infty$ for given $z$, whereas an asymptotic series tends to $f(z)$ when $z \to \infty$ for given $n$. Asymptotic expansions can be used when $z$ is complex, but in most cases the expansion will be valid only over a limited range of arg $z$.

It may happen that a function $f(z)$ does not possess an asymptotic expansion like (3.12), but that there is a function $g(z)$ such that

$$f(z)/g(z) \sim A_0 + A_1/z + A_2/z^2 + \cdots \quad \text{as } z \to \infty$$

In this case one writes

$$f(z) \sim g(z)(A_0 + A_1/z + A_2/z^2 + \cdots) \quad \text{as } z \to \infty \tag{3.13}$$

A familiar example of (3.13) is the asymptotic expansion of the gamma function

$$\Gamma(z) \sim e^{-z} z^{z-\frac{1}{2}} (2\pi)^{\frac{1}{2}} [1 + 1/12z + \cdots]$$

This series gives a better approximation to $\Gamma(z)$ for large $z$ than for small $z$, but the leading term is accurate to within 10% even for $z = 1$.

The relations (3.12) and (3.13) are asymptotic approximations valid for large $z$. Asymptotic series with a small parameter also occur. For example, Euler's series

$$S_n(x) = 1 - x + 2!x^2 - 3!x^3 + \cdots + (-1)^n n! x^n$$

is an asymptotic approximation for the integral

$$f(x) = \int_0^\infty e^{-t}/(1 + xt)dt \quad \text{as } x \to 0$$

In this case for any $n$

$$(f(x) - S_n(x))/x^n \to 0 \quad \text{as } x \to 0$$

We conclude this section by quoting asymptotic expansions for the Airy function Ai($z$). These are used to obtain the WKB connection formula in section 3.3. One expansion is (Abramowitz & Stegun, 1965, 10.4.59)

$$\text{Ai}(z) \sim \tfrac{1}{2}\pi^{-\frac{1}{2}} z^{-\frac{1}{4}} \exp(-\tfrac{2}{3} z^{\frac{3}{2}})(1 - (5/48)z^{-\frac{3}{2}} + \cdots) \tag{3.14}$$

which is valid for large complex $z$ provided $-\pi < \arg z < \pi$. This is an asymptotic expansion in the sense of (3.13) in powers of $z^{-\frac{3}{2}}$. Only the leading term of (3.14) is used for the applications considered in this book. When arg $z = \pi$, that is, when $z$ is real and negative, the expansion (3.14) fails. For such values of $z$ there is another expansion which is easiest to

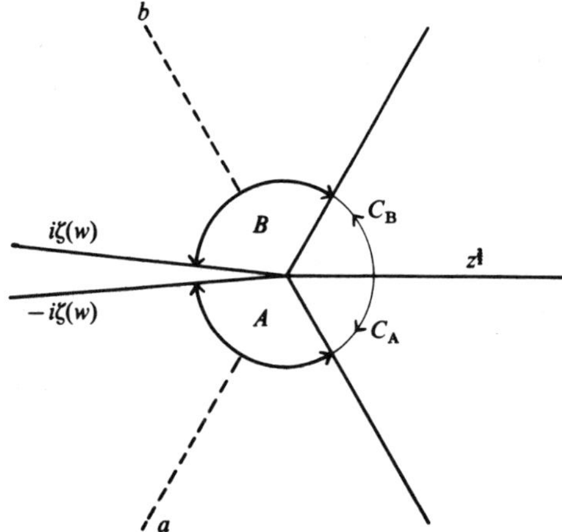


Fig. 3.1 Regions of validity of asymptotic approximations of the Airy function. The term in equation (3.16) proportional to $\exp(-i\zeta(w))$ is dominant in the region $A$, while the term proportional to $\exp(i\zeta(w))$ is dominant in the region $B$.

write in terms of $w = -z$. The leading term is

$$\left.\begin{aligned} \mathrm{Ai}(-w) &\sim \pi^{-\frac{1}{2}} w^{-\frac{1}{4}} \sin(\zeta(w) + \tfrac{1}{4}\pi) \\ \zeta(w) &= \tfrac{2}{3} w^{\frac{3}{2}} \end{aligned}\right\} \tag{3.15}$$

This approximation holds for large complex $w$ provided $-2\pi/3 < \arg w < 2\pi/3$.

The relation between the approximations (3.14) and (3.15) can be understood by writing the sine function in (3.15) in terms of complex exponentials

$$\mathrm{Ai}(-w) \sim \tfrac{1}{2}\pi^{-\frac{1}{2}} w^{-\frac{1}{4}} [\exp(i\zeta(w) - \tfrac{1}{4}i\pi) + \exp(-i\zeta(w) + \tfrac{1}{4}i\pi)] \tag{3.16}$$

Points in the region $A$ in fig. 3.1 satisfy the conditions $-\pi < \arg z < -\pi/3$ or $0 < \arg w < 2\pi/3$. The term proportional to $\exp(-i\zeta(w))$ increases exponentially as $|w| \to \infty$ and is said to be exponentially dominant. The other term in (3.16) proportional to $\exp(i\zeta(w))$ is exponentially recessive and is negligible when $|w|$ is large. In the same region $z = w\exp(-i\pi)$ and

$$\begin{aligned} z^{-\frac{1}{4}} &= w^{-\frac{1}{4}} \exp(\tfrac{1}{4}i\pi) \\ \tfrac{2}{3} z^{\frac{3}{2}} &= \tfrac{2}{3} w^{\frac{3}{2}} \exp(-\tfrac{3}{2}i\pi) = i\zeta(w) \end{aligned}$$

so that the asymptotic form (3.14) coincides with the exponentially dominant term in (3.16). In the region $B$ in fig. 3.1 $\pi/3 < \arg z < \pi$ and $-2\pi/3 < \arg w < 0$. In this region $z = w\exp(i\pi)$ the term in (3.16) proportional to $\exp(i\zeta(w))$ is exponentially dominant and it is equal to (3.14).

There is another way to look at this same argument. The Airy function

Ai($z$) is an analytic function of $z$ for all $z$. The asymptotic forms (3.14) and (3.16) have a branch point at $z=0$ and give a good approximation to Ai($z$) only in certain regions of the complex plane. Making an analytic continuation of (3.14) along the curve $C_A$ in the complex plane (fig. 3.1) gives the term proportional to $\exp(-i\zeta(w))$ in (3.16), while an analytic continuation along $C_B$ gives the term proportional to $\exp(i\zeta(w))$.

The function $z^{\frac{3}{2}}$ is real on the real $z$-axis for $z>0$ and on the dashed lines $a$ and $b$ in fig. 3.1. These lines are called Stokes lines. They are important in the derivation of the WKB connection formula at the end of section 3.3.

## 3.3 Connection formulae

Consider a particle with energy $E$ moving in the potential shown in fig. 3.2. The region $b>x>a$ is classically accessible ($E>V(x)$) while the regions $x<a$ and $x>b$ are classically forbidden ($E<V(x)$). Suppose we seek a solution which decays exponentially as $x\to-\infty$. According to (3.5) the WKB wave function for $x<a$ is then

$$\psi(x)\simeq\tfrac{1}{2}C\gamma^{-\frac{1}{2}}\exp\left(-\int_x^a\gamma(x)dx\right);\quad \gamma(x)>0 \tag{3.17}$$

The solution for $x>a$ is oscillatory and can be written as

$$\begin{aligned}\psi(x)&\simeq Ak^{-\frac{1}{2}}\sin\left(\int_a^x k(x)dx+\alpha\right)\\&=k^{-\frac{1}{2}}\left(C_1\exp\left(i\int_a^x k(x)dx\right)+C_2\exp\left(-i\int_a^x k(x)dx\right)\right)\end{aligned} \tag{3.18}$$

Near $x=a$ both $k(x)$ and $\gamma(x)$ tend to zero, the condition (3.6) for the validity of the WKB method is not satisfied, and the solutions (3.17) and (3.18) cannot be used. It is necessary to find a procedure to connect the solutions on each side of the turning point. This connection problem has been discussed by many authors and derivations are given in most textbooks on quantum mechanics. Dingle (1973) has given an interesting critical discussion of the problem and an account of its history. Here we first state the result and then outline a derivation.

At the left turning point at $x=a$

$$\left.\begin{aligned}\psi(x)&\simeq\tfrac{1}{2}C\gamma^{-\frac{1}{2}}\exp\left(-\int_x^a\gamma(x)dx\right)\quad\text{for } x<a\\ \psi(x)&\simeq Ck^{-\frac{1}{2}}\sin\left(\int_a^x k(x)dx+\tfrac{1}{4}\pi\right)\quad\text{for } x>a\end{aligned}\right\} \tag{3.19}$$

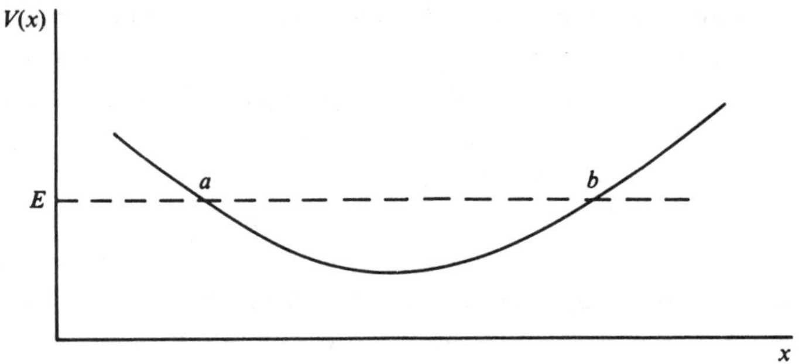


Fig. 3.2. Sketch of a potential $V(x)$ showing classical turning points ($V(x) = E$) at $x = a$ and $x = b$.

while at the right turning point at $x = b$

$$\left.\begin{aligned} \psi(x) &= \tfrac{1}{2}C'\gamma^{-\frac{1}{2}}\exp\left(-\int_b^x \gamma(x)dx\right) \quad \text{for } x > b \\ \psi(x) &\simeq C'k^{-\frac{1}{2}}\sin\left(\int_x^b k(x)dx + \tfrac{1}{4}\pi\right) \quad \text{for } x < b \end{aligned}\right\} \tag{3.20}$$

To derive the matching condition at the left turning point we make a linear approximation to the potential near $x = a$

$$k^2(x) = (2m/\hbar^2)(E - V(x)) \simeq \beta^3(x - a)$$

with

$$\beta^3 = (2m/\hbar^2)|V'(a)|$$

When $x < a$

$$\int_x^a \gamma(x)dx \simeq \tfrac{2}{3}\{\beta(a - x)\}^{\frac{3}{2}} = \xi(x)$$

while for $x > a$

$$\int_a^x k(x)dx \simeq \tfrac{2}{3}\{\beta(x - a)\}^{\frac{3}{2}} = \zeta(x)$$

For $x$ near $a$ the Schrödinger equation becomes

$$d^2\psi/dx^2 + \beta^3(x - a)\psi = 0 \tag{3.21}$$

and has an exact solution in terms of the Airy function

$$\psi(x) = \text{Ai}(-\beta(x - a)) \tag{3.22}$$

When $x - a$ is sufficiently large and negative the asymptotic form

(3.14) of the Airy function gives

$$\psi(x) \sim \tfrac{1}{2}\pi^{-\frac{1}{2}}\{\beta(a-x)\}^{-\frac{1}{4}}\exp(-\xi(x))$$

which agrees with (3.17) if $C=(\beta/\pi)^{\frac{1}{2}}$. When $x-a$ is large and positive the second asymptotic form (3.15) of the Airy function gives

$$\psi(x) \sim \pi^{-\frac{1}{2}}\{\beta(x-a)\}^{-\frac{1}{4}}\sin(\zeta(x)+\tfrac{1}{4}\pi)$$

This agrees with (3.18) if $A=C$ and $\alpha=\frac{1}{4}\pi$, and is just the connection formula (3.19). The connection formula (3.20) at $x=b$ can be derived in a similar way.

Equations (3.19) and (3.20) are connection formulae relating exponentially decaying solutions in the classically inaccessible regions with oscillating solutions in the allowed regions. There are similar formulae connecting exponentially increasing solutions in the forbidden regions with oscillating solutions in the allowed regions. They can be obtained by the method described above by using the second solution Bi($z$) of Airy's equation. We write these connection formulae for the left turning point at $x=a$ as

$$\left.\begin{aligned}\psi(x) &\simeq D\gamma^{-\frac{1}{2}}\exp\left(\int_x^a \gamma(x)dx\right) \quad \text{for } x<a\\ \psi(x) &\simeq D\gamma^{-\frac{1}{2}}\cos\left(\int_a^x k(x)+\tfrac{1}{4}\pi\right)\\ &\quad + E\gamma^{-\frac{1}{2}}\sin\left(\int_a^x k(x)dx+\tfrac{1}{4}\pi\right) \quad \text{for } x>a\end{aligned}\right\} \tag{3.23}$$

There is considerable discussion in the literature about whether the coefficient $E$ can be determined. This is because the solution (3.23) is exponentially large for $x<a$. Adding a component of the decaying solution (3.17), (3.18) makes hardly any difference for $x<a$ since (3.17) is exponentially small there, but it does change the value of the coefficient $E$. Dingle (1973) analyses this discussion and concludes that $E=0$ if there is (exactly) no exponentially decreasing solution for $x<a$. On the other hand, if the exponentially increasing solution for $x<a$ is suspected to be contaminated by an exponentially decreasing component of unknown magnitude, nothing can be said about the value of $E$. There is a similar connection formula for the right turning point at $x=b$.

The connection formulae (3.19) and (3.20) can also be obtained by an analytic continuation technique. (Furry, 1947). We mention it here because it is a useful method and because it is the basis of an important work by Knoll & Schaeffer (1976) on applications of semi-classical methods to

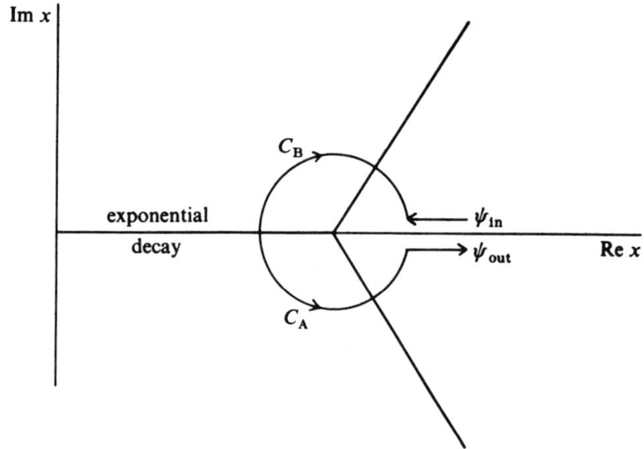


Fig. 3.3. Connection formulae by analytic continuation in the complex $x$-plane. The wave function $\psi(x)$ decays exponentially for $x<a$. Analytic continuation along $C_A$ gives an outgoing wave $\psi_{out}$ for $x>a$. Analytic continuation along $C_B$ gives an incoming wave $\psi_{in}$.

nuclear scattering problems. Figure 3.3 shows the complex $x$-plane near the left turning point at $x=a$. Introduce the notation

$$\xi(x) = -\int_a^x \gamma(x)dx = \int_x^a \gamma(x)dx \tag{3.24}$$

If the potential $V(x)$ is an analytical function of $x$ in some region around $x=a$ then both $\gamma(x)$ and $\xi(x)$ are analytical functions of $x$ in the same region but with branch points at $x=a$. If $E$ is real and $V(x)$ is real for real $x$ then $\xi(x)$ is real and positive along the real axis for $x<a$. The function $\xi(x)$ is also real along two other lines, $B$ and $C$, radiating from $x=a$. The three lines along which $\xi(x)$ is real are called Stokes lines.

If $\xi(x)$ is analytically continued to the real axis for $x>a$ avoiding the branch point at $x=a$ by passing it in the lower half of the $x$-plane then its phase changes by $3\pi/2$ so that

$$\xi(x) = |\xi(x)|\exp(\tfrac{3}{2}i\pi) = -i|\xi(x)| \quad \text{for } x>a$$

where

$$|\xi(x)| = \int_a^x k(x)dx \quad \text{for } x>a$$

Similarly

$$\gamma(x) = |\gamma(x)|\exp(\tfrac{1}{2}i\pi) = i|\gamma(x)| = ik(x)$$

Making an analytic continuation in the upper half of the $x$-plane gives a phase change of $-3\pi/2$ for $\xi$ and $-\frac{1}{2}\pi$ for $\gamma$ and

$$\xi(x) = i|\xi(x)|, \quad \gamma(x) = -i|\gamma(x)| \quad \text{for } x > a$$

Suppose we have the WKB solution (3.17) on the real axis for $x < a$

$$\psi(x) \simeq \tfrac{1}{2}C\gamma^{-\frac{1}{2}}\exp(-\xi(x)) \tag{3.25}$$

Then the solution can be found for $x > a$ by analytic continuation using the principle of exponential dominance (Stokes, 1904). Crossing a Stokes line the exponentially dominant term is continuous but the subdominant term has a discontinuity in the presence of a dominant term. Otherwise it too is continuous.

The solution (3.25) for $x < a$ is subdominant but the dominant exponentially increasing solution is zero. Hence (3.25) can be continued off the real axis. Continuing in the lower half-plane the solution (3.25) becomes dominant on the Stokes line $C$ and can be continued past it to the real axis. There it corresponds to an out-going wave

$$\tfrac{1}{2}C|\gamma|^{-\frac{1}{2}}\exp(i|\xi(x)|)\exp(-\tfrac{1}{4}i\pi)$$

Making an analytic continuation in the upper half-plane gives an incoming wave for $x > a$

$$\tfrac{1}{2}C|\gamma|^{-\frac{1}{2}}\exp(-i|\xi(x)| + \tfrac{1}{4}i\pi)$$

Hence the total solution for $x > a$ is

$$\psi(x) \simeq \tfrac{1}{2}|\gamma|^{-\frac{1}{2}}(C_{\text{out}}\exp(i|\xi(x)|) + C_{\text{in}}\exp(-i|\xi(x)|))$$

with $C_{\text{out}} = C\exp(-\frac{1}{4}i\pi)$ and $C_{\text{in}} = C\exp(\frac{1}{4}i\pi)$. This result can also be written as

$$C = \exp(-\tfrac{1}{4}i\pi)C_{\text{in}}, \quad C_{\text{out}} = \exp(-\tfrac{1}{2}i\pi)C_{\text{in}} \tag{3.26}$$

and is equivalent to the connection formula (3.19). Figure 3.3 and equation (3.19) can be used to give connection formulae at any isolated turning point, real or complex. It is only necessary to rotate the diagram to the correct orientation. The analytic properties of the WKB solutions discussed here are similar to the analytic properties of the asymptotic form of the Airy function which were analysed in section 3.2 and illustrated in fig. 3.1.

## 3.4 Complex turning points

As discussed in section 2.2 heavy-ion scattering is often analysed using an optical potential which is complex. Its imaginary part $\text{Im}\,V(x) = -W(x)$ is negative and describes absorption of flux from the elastic scattering channel by various reactions.

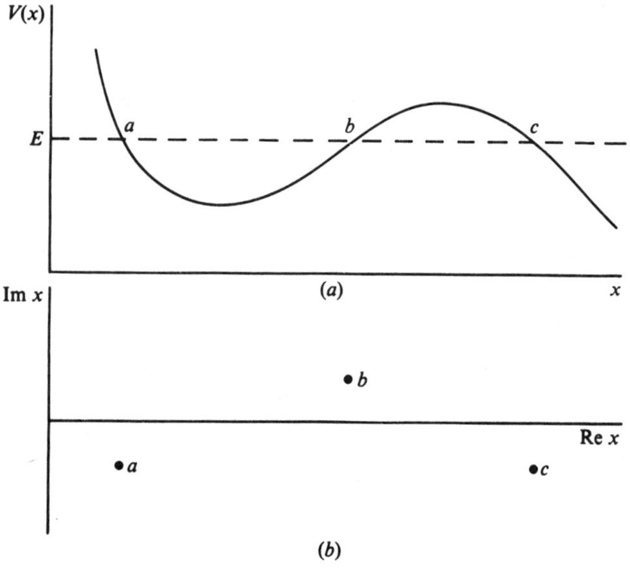


Fig. 3.4. (*a*) Sketch of a potential $V(x)$ with three turning points ($V(x)=E$) at $x=a$, $b$ and $c$. (*b*) Positions of the turning points in the complex $x$-plane if $V(x)$ is complex and $\operatorname{Im} V(x)<0$.

The WKB wave functions (3.17) and (3.18) can still be used when $V(x)$ is complex. The only change is that $k(x)$ and $\gamma(x)$ are also complex. The situation is more complicated when there are turning points. A turning point is a solution of the equation

$$V(x)-E=0 \tag{3.27}$$

If $V(x)$ is complex the solutions of (3.27) give complex turning points. When $\operatorname{Im} V(x)$ is small an approximate solution of (3.27) can be found in the following way:

Write $V(x)=U(x)-iW(x)$ and let $x_0$ be a real solution of

$$U(x)-E=0.$$

Then to first order in $W(x)$ the complex turning point is at $x_1$ where

$$x_1=x_0+iW(x_0)/V'(x_0) \tag{3.28}$$

At a left turning point like the one at $x=a$ in fig. 3.2 we have the derivative $V'(x)<0$. As $W(x_0)$ is positive $\operatorname{Im} x_1<0$, and the turning point moves into the lower part of the complex $x$-plane. At a right turning point $\operatorname{Im} x_1>0$ and the turning point moves into the upper part of the complex $x$-plane.

Figure 3.4 shows a typical real potential with three turning points at $a$, $b$ and $c$. The lower part of the diagram shows the position of the corresponding complex turning points when a small negative imaginary part is added to $V(x)$. The local wave number $k(x)$ is proportional to the square root of $E - V(x)$ and has branch points at the turning points of $V(x)$. Normally it is convenient to choose the corresponding branch cuts so that $V(x)$ is an analytic function of $x$ on and near the real axis.

Complex turning points can occur even if $V(x)$ is a real potential. Consider the problem of a potential barrier (fig. 3.4). If the incident energy is below the top of the barrier there are two real turning points at $b$ and $c$. When the incident energy is above the top of the barrier there is a complex conjugate pair of solutions of (3.27). Hence, there are two complex turning points, that is, one with $\operatorname{Im} x_1 < 0$ and the other with $\operatorname{Im} x_2 > 0$. Classically a particle is not reflected if its incident energy is above the top of the potential barrier. In a full quantal theory there is a reflected wave, and in the semi-classical theory this reflected wave is associated with the complex turning points.

## 3.5 Reflection by an exponential potential

In this section we discuss an example of a problem with a complex turning point to illustrate the results obtained in previous sections. The example is also of interest for heavy-ion scattering problems. Consider the motion of a particle with energy $E$ in a potential

$$V(x) = V_0 \exp(-x/a) \tag{3.29}$$

To begin with, $V_0$ will be chosen to be real and positive, but later the problem with a complex $V_0$ will be discussed.

The local momentum corresponding to the potential (3.29) is given by

$$k^2(x) = k^2[1 - (V_0/E)\exp(-x/a)] \tag{3.30}$$

with $k^2 = 2mE/\hbar^2$. There is a left-hand turning point ($k(x) = 0$) at

$$x_0 = a\ln(V_0/E) \tag{3.31}$$

The WKB solution requires the evaluation of the integral

$$w(x, x_0) = \int_{x_0}^{x} k(x)dx = k\int_{x_0}^{x} [1 - (V_0/E)\exp(-x/a)]^{\frac{1}{2}}dx$$

If $x - x_0 \gg a$ the integral can be calculated and gives

$$\begin{aligned} w(x, x_0) &\sim k(x - x_0) - ka(1 - \ln 2) \\ &= k(x - x_0) - 0.307\,ka \end{aligned} \tag{3.32}$$

and the WKB solution can be written as

$$\psi(x) \sim C_1 \exp(ikx) + C_2 \exp(-ikx) \tag{3.33}$$

where the WKB matching condition (3.26) gives

$$\left.\begin{aligned} C_1 &= C \exp(-i\phi(x_0, k)) \\ C_2 &= C \exp(i\phi(x_0, k)) \end{aligned}\right\} \tag{3.34}$$

$$\phi(x_0, k) = kx_0 - 0.307\, ka + \tfrac{1}{4}\pi \tag{3.35}$$

The Schrödinger equation with the potential (3.29) has an exact solution in terms of a modified Bessel function $K_\nu(z)$, (Abramowitz & Stegun, 1965, p. 375),

$$\psi(x) = K_{2ika}(2ka \exp((x_0 - x)/2a))$$

This exact solution has the asymptotic form (3.33) for large $x$ with a phase

$$\phi = kx_0 + 2ka \ln ka - \operatorname{Im} \Gamma(1 + 2ika) \tag{3.36}$$

If $ka \gg 1$ the gamma function in (3.36) can be replaced by its asymptotic form and $\phi$ reduces to the WKB result (3.35). Thus in this example we have an explicit condition for the validity of the WKB and the expansion parameter for corrections is $(ka)^{-1}$. The first term neglected in the WKB result for $\phi$ is $(24\, ka)^{-1}$.

Now we consider the case where $V_0$ is complex and put

$$V_0 = |V_0| \exp(-i\alpha) \tag{3.37}$$

If $0 < \alpha < \pi$ then $\operatorname{Im} V_0 < 0$ as it should be for an absorbing potential while $\alpha = \pi$ gives an attractive real potential. The case $\alpha = \frac{1}{2}\pi$ is a purely imaginary potential. Equation (3.31) now gives a complex turning point with

$$\operatorname{Im} x_0 = -a\alpha < 0 \tag{3.38}$$

In the solution (3.33) the term proportional to $C_2$ represents a wave incident on the potential from the right and the term proportional to $C_1$ is a reflected wave. The reflection coefficient calculated from (3.34) is

$$|C_1/C_2|^2 = \exp(4k \operatorname{Im} x_0) = \exp(-4k\alpha a) \tag{3.39}$$

## 3.6 The radial Schrödinger equation

Elastic scattering phase shifts for a partial wave $l$ are obtained by solving the Schrödinger equation

$$\frac{d^2}{dr^2} F_l + k^2(r) F_l = 0$$

where

$$k^2(r) = (2\mu/\hbar^2)(E - V_{\text{eff}}(r))$$
$$V_{\text{eff}}(r) = V_n(r) + V_C(r) + \hbar^2 l(l+1)/2\mu r^2 \tag{3.40}$$

$V_n(r)$ is the nuclear potential acting between the target and the projectile and $V_C(r)$ is the Coulomb interaction. For large $r$ the wave function $F_l(r)$ has the asymptotic form

$$F_l(r) \sim \sin(kr - n\ln 2kr - \tfrac{1}{2}l\pi + \sigma_l + \delta_{nl}) \tag{3.41}$$

Here $k$ is the asymptotic wave number ($k = (2\mu E/\hbar^2)^{\frac{1}{2}}$), $\sigma_l$ is the Coulomb phase shift, $\delta_{nl}$ is the nuclear phase shift and $n$ is Sommerfeld's parameter.

Suppose that there is one classical turning point at $r_0$

$$k(r_0) = 0, \quad V_{\text{eff}}(r_0) = E$$

and that the region $r > r_0$ is allowed classically ($E > V_{\text{eff}}(r)$) and $r < r_0$ is classically forbidden ($E < V_{\text{eff}}(r)$). According to section 3.3 the WKB wave function in the classically allowed region is

$$F_l(r) \simeq (k(r))^{-\frac{1}{2}} \sin\left(\int_{r_0}^{r} k(r)dr + \tfrac{1}{4}\pi\right) \tag{3.42}$$

The phase shifts can be found by combining (3.41) and (3.42). The matching condition at $r_0$ used in (3.42) was obtained in section 3.3 by assuming that $r_0$ is an isolated turning point. However, the argument must be modified when using the WKB method to solve (3.40) for small $l$, because of the singularity at $r = 0$ in the centrifugal potential. It can be shown that the quantity $l(l+1)$ in (3.40) should be replaced by $(l+\frac{1}{2})^2$ if the WKB method is to be used. This modification is known as the Langer correction (Langer, 1934, 1937). It is unimportant for large $l$, but gives a result which is a valid approximation for all $l$.

To obtain the Langer correction we change both the independent variables in the Schrödinger equation and put $r = e^x$ and $F = G\exp(\frac{1}{2}x)$. The point $r = 0$ is transformed into $x = -\infty$ and the singularity is moved out of harm's way. In the new variables the Schrödinger equation is

$$\frac{d^2G}{dx^2} + \left[\frac{2\mu}{\hbar^2}(E - V)e^{2x} - (l+\tfrac{1}{2})^2\right]G = 0$$

This equation no longer has any singularity, but it does have a simple isolated turning point at $x_0$, where the coefficient of $G$ vanishes. The simple matching condition (3.19) can be used in the variable $x$ and the resulting expression transformed back to the variable $r$. This leads to the formula (3.42) for the wave function with $l(l+1)$ replaced by $(l+\frac{1}{2})^2$. The turning point $r_0$ is now the point where the modified $k(r_0) = 0$.

The case where $l = 0$ merits a special discussion. Consider the case where $k_0^2(r) = (2\mu/\hbar^2)(E - V(r)) > 0$ for $r > 0$. The radial wave function $F_0(r)$ should satisfy the boundary condition $F_0(0) = 0$ and the WKB wave function is

$$F_0(r) = (k_0(r))^{-\frac{1}{2}} \sin\left(\int_0^r k_0(r)dr\right) \tag{3.43}$$

Another approach is to use the formula (3.42) with the Langer modification. When $l = 0$ we have $(l + \frac{1}{2})^2 = \frac{1}{4}$ and (3.42) gives

$$F_0(r) = (k_0^2 - \tfrac{1}{4}r^2)^{-\frac{1}{4}} \sin\left(\int_{r_0}^r (k_0^2(r) - \tfrac{1}{4}r^2)^{\frac{1}{2}} dr + \tfrac{1}{4}\pi\right) \tag{3.44}$$

Although the expressions (3.43) and (3.44) look very different they are numerically almost the same for large $r$. This can be seen explicitly when $k_0$ is constant and the integrals can be evaluated. The argument of the sine function in (3.44) is

$$(k_0^2 r^2 - \tfrac{1}{4})^{\frac{1}{2}} + \tfrac{1}{2}\sin^{-1}(1/(2k_0 r)) \sim k_0 r + 1/(8k_0 r) + \cdots$$

Thus it is consistent to use Langer's modification even in the case $l = 0$, which is interesting because $l = 0$ does not have to be considered as a special case.

## 3.7 Coulomb scattering

With the Langer correction the semi-classical form for a radial scattering wave function for a point Coulomb field is given by

$$F_l(r) = (k_C(r))^{-\frac{1}{2}} \sin\left(\int_{r_C}^r k_C(r)dr + \tfrac{1}{4}\pi\right)$$

where

$$k_C(r) = k[1 - 2a_C/r - b^2/r^2]^{\frac{1}{2}} \tag{3.45}$$

In (3.45) $a_C$ is the Coulomb length parameter, $b$ is the impact parameter ($kb = \lambda = l + \frac{1}{2}$) and $r_C$ is the Coulomb turning point ($k_C(r_C) = 0$). The integral over $k_C(r)$ can be evaluated to give

$$\int_{r_C}^r k_C(r)dr = rk_C(r) - n\ln(rk_C(r) + rk - n) + \tfrac{1}{2}n\ln(n^2 + \lambda^2)$$
$$+ \lambda\sin^{-1}\left[\frac{nkr + \lambda^2}{kr(n^2 + \lambda^2)^{\frac{1}{2}}}\right] - \tfrac{1}{2}\pi\lambda \tag{3.46}$$

When $r$ is large the Coulomb wave function has exactly the asymptotic form (3.41) with Coulomb phase

$$\sigma_l = \sigma(\lambda) = \tfrac{1}{2} n \ln(n^2 + \lambda^2) - n + \lambda \tan^{-1}(n/\lambda) \tag{3.47}$$

The exact expression for the Coulomb phase is (Messiah, 1959)

$$\sigma_l = \arg \Gamma(l + 1 + in)$$

which is asymptotically equal to (3.47) if either $n \gg 1$ or $\lambda^2 \gg n$.

## 3.8 Phase shifts and deflection functions

The WKB wave function (3.42) contains the integral

$$\int_{r_0}^{r} k(r)dr = \int_{r_C}^{r} k_C(r)dr + \left[ \int_{r_0}^{r} k(r)dr - \int_{r_C}^{r} k_C(r)dr \right] \tag{3.48}$$

The first term in (3.48) is the Coulomb WKB integral (3.46). If $r$ is outside the range of the nuclear force $V_n(r)$ then $k(r) = k_C(r)$ and the term in square brackets in (3.48) is independent of the upper limit of integration and is equal to the nuclear phase shift in (3.41). Hence, the WKB formula for the nuclear phase is

$$\delta_n(\lambda, E) \simeq \int_{r_0}^{R} k(r)dr - \int_{r_C}^{R} k_C(r)dr \tag{3.49}$$

In this formula $k(r)$ is given by

$$k^2(r) = (2\mu/\hbar^2)(E - V_{\text{eff}}(r))$$
$$V_{\text{eff}}(r) = V_n(r) + V_C(r) + \hbar^2\lambda^2/(2\mu r^2) \tag{3.50}$$

where $\lambda = l + \frac{1}{2}$ (see section 3.6) and $k_C(r)$ by the same formula with $V_n = 0$. The classical turning points $r_o$ and $r_C$ are defined by $k(r) = 0$ and $k_C(r_C) = 0$ respectively. The upper limit $R$ of the integrals is chosen to be outside the range of the nuclear interaction so that $k(r) = k_C(r)$ for $r \geqslant R$.

Equation (3.49) is the correct semi-classical formula for the nuclear phase if there is only one classical turning point. Knoll & Schaeffer (1976) and Landowne *et al.* (1976) have shown that it is valid even if the nuclear potential $V_n(r)$ is complex, provided $r_0$ is the appropriate complex turning point (i.e. the complex solution of $k(r_0) = 0$), and the integral is taken along a suitable path in the complex $r$-plane. The extension of (3.49), when there are several complex turning points, has also been discussed by Knoll & Schaeffer (1976). We discuss a three-turning-point problem in chapter 6.

In the case of a real potential there is a close connection between the WKB phase (3.49) and quantities characterizing classical elastic scattering. The product $p(r) = \hbar k(r)$ is the radial component of the relative momentum of the target and projectile and is related to the radial component of relative velocity by

$$p(r) = \hbar k(r) = \mu \dot{r}$$

Differentiating (3.49) gives

$$\begin{aligned}\partial\delta_n/\partial E &= (\mu/\hbar^2)\left[\int_{r_0}^{R} dr/k(r) - \int_{r_C}^{R} dr/k_C(r)\right]\\ &= (1/\hbar)\left[\int_{r_0}^{R} dr/\dot{r} - \int_{r_C}^{R} dr/\dot{r}_C\right]\\ &= (1/2\hbar)[T(R) - T_C(R)] \end{aligned} \tag{3.51}$$

In (3.51) $T(R)$ is the classical transit time from a radial separation $R$ before the collision to a separation $R$ after the collision, and $T_C(R)$ is the corresponding quantity for a Coulomb orbit. Thus, (3.51) gives the modification of the collision time produced by the potential $V_n(r)$. In the same way

$$\begin{aligned}\partial\delta_n/\partial\lambda &= -\lambda\left[\int_{r_0}^{R} dr/(r^2 k(r)) - \int_{r_C}^{R} dr/(r^2 k_C(r))\right]\\ &= -(\hbar\lambda/\mu)\left[\int_{r_0}^{R} dr/(r^2\dot{r}) - \int_{r_C}^{R} dr/(r^2\dot{r}_C)\right]\end{aligned}$$

Using the classical formula for the angular momentum $\hbar\lambda = \mu r^2\dot{\theta}$ one obtains

$$\partial\delta_n/\partial\lambda = \tfrac{1}{2}[\Theta(\lambda, E) - \Theta_C(\lambda, E)] \tag{3.52}$$

Here $\Theta(\lambda, E)$ and $\Theta_C(\lambda, E)$ are the classical scattering angles with and without the nuclear interaction. The integrals converge for large $R$ and the upper limit can be replaced by infinity. The semi-classical formula (3.47) for the Coulomb phase gives

$$\partial\sigma(\lambda)/\partial\lambda = \cot^{-1}(\lambda/n) = \tfrac{1}{2}\Theta_C(\lambda, E)$$

Hence, defining the total phase $\delta(\lambda) = \sigma(\lambda) + \delta_n^{(\lambda)}$ we obtain

$$\partial\delta/\partial\lambda = \tfrac{1}{2}\Theta(\lambda, E) = \tfrac{1}{2}\pi - \lambda\int_{r_0}^{\infty} dr/(r^2 k(r)) \tag{3.53}$$

so that the derivative of the total phase is half the classical scattering angle.

It is possible for classical paths to encircle the origin before emerging. Thus the magnitude $|\Theta|$ of the classical scattering angle may exceed $\pi$, unlike the observation angle $\theta$ which lies between 0 and $\pi$. Figure 3.5 illustrates a collection of orbits, all of which correspond to the same

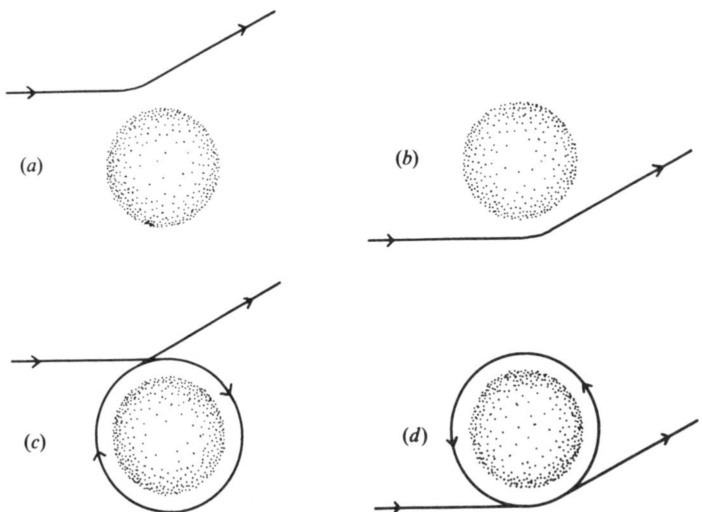


Fig. 3.5. Classical trajectories with the same angle of observation $\theta$ but different scattering angles $\Theta(\lambda)$: (a) $\Theta(\lambda_1)=\theta$, (b) $\Theta(\lambda_2)=-\theta$, (c) $\Theta(\lambda_3)=\theta-2\pi$ and (d) $\Theta(\lambda_4)=-\theta-2\pi$. The cases (a) and (c) are nearside trajectories while (b) and (d) are farside.

angle of observation $\theta$. The scattering angles are

$$\Theta(\lambda_1)=\theta, \qquad \Theta(\lambda_2)=-\theta$$
$$\Theta(\lambda_3)=\theta-2\pi, \quad \Theta(\lambda_4)=-\theta-2\pi$$

Taking cylindrical symmetry into account the classical scattering angles $\Theta$ which contribute to an observation angle $\theta$ must satisfy

$$\Theta=\pm\theta-2m\pi; \quad m=0,1,2,\ldots \tag{3.54}$$

The integer $m$ cannot be negative because values of $\Theta$ exceeding $\pi$ cannot occur classically.

## 3.9 Structure of the nuclear phase

The results obtained in the previous sections can be used to understand the relation between the optical potential, the phase shifts $\delta_l$ and the partial wave amplitudes $S_l$. The radial Schrödinger equation contains an effective potential (3.40)

$$V_{\text{eff}}(l,r)=V_{\text{n}}(r)+V_{\text{C}}(r)+\hbar^2 l(l+1)/(2\mu r^2)$$

which is a sum of nuclear, Coulomb and centrifugal components. Figure 3.6 shows Re $V_{\text{eff}}$ plotted against $r$ for several $l$ values for a typical

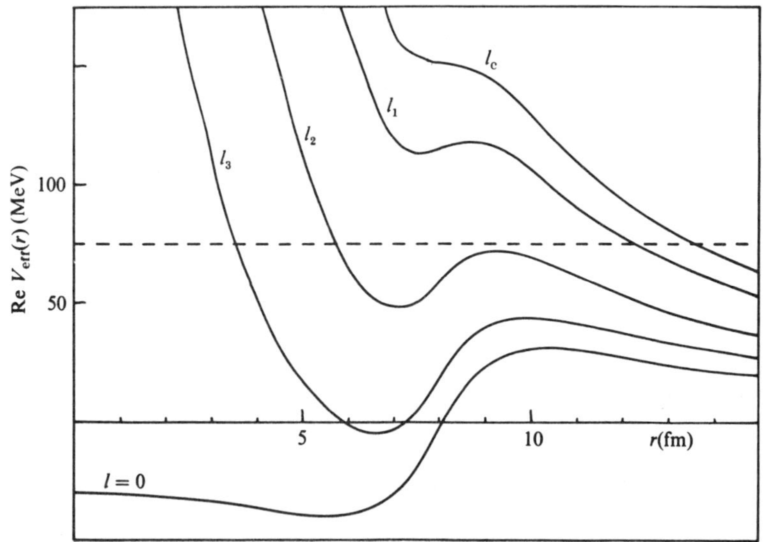


Fig. 3.6. Plots of Re $V_{eff}(r)$ for several values of angular momentum $l$ for a Woods–Saxon potential with $V_0 = 90$ MeV, range $R = 7.92$ fm and surface diffuseness $a = 0.5$ fm. The values of $l$ are $l_1 = 70$, $l_2 = 50$, $l_3 = 30$, $l_C = 80$. The parameters are appropriate for scattering of $^{18}O$ by $^{58}Ni$.

optical potential. In this example the total potential has a barrier whose height $V_B(l)$ increases with $l$. The horizontal line denotes the incident centre-of-mass energy $E$. We need to consider three different angular momentum regions in order to understand the variation of $S_l$ with $l$.

I. $V_B(l_1)$ larger than $E$: This is the high angular momentum region. The separation of the target and projectile and the classical turning point $A$ are large and the nuclear part of the optical potential is small in the classically allowed region beyond $A$. The magnitude of $S_l$ is near unity and the nuclear phase shift is small.

II. $V_B(l_2)$ near $E$: This is a band of angular momenta near the grazing angular momentum $l_g$. Many factors influence $S_l$ in this region. The nuclear part of the optical potential has an important influence on the scattering, and absorption by the imaginary part $W(r)$ of the optical potential can be significant. The incident energy is near the top of the barrier and it may be necessary to consider the effects of barrier penetration. In a classical system the orbiting phenomenon occurs in this angular momentum range and the classical orbiting angular momentum is determined by the condition

$$V_B(l_0) = E \tag{3.55}$$

Orbiting happens because the radial motion is very slow in the barrier region and the projectile can circle around the target several times before separating from it. The importance of barrier penetration and orbiting depends on the strength of the imaginary potential $W_B$ at the barrier position. If $W_B$ is strong enough the incident wave is absorbed outside the barrier and barrier penetration or orbiting do not occur.

The following argument gives a measure of the importance of absorption. Suppose that the relative position of target and projectile remains in the barrier region for a time $\tau$. The argument at the beginning of section 2.2 shows that $2W/\hbar$ is the probability per unit time for absorption out of the elastic channel. Hence, an orbit for which the absorption probability is one-half has

$$\tau_{\frac{1}{2}} \approx (\tfrac{1}{2}\ln 2)\hbar/W_B \approx 0.35\hbar/W_B$$

The deflection of the orbit by the nuclear force in the same time is $\Delta\Theta \approx F_{nB}\tau/p_0$ where $p_0 = \hbar k_0$ is the tangential momentum, $F_{nB} \approx U_B/a$ the nuclear force strength, $U_B$ the real part of the optical potential at the barrier position and $a$ is the surface diffuseness of the nuclear potential. Hence, the deflection angle in a time $\tau_{\frac{1}{2}}$ is

$$\Delta\Theta_{\frac{1}{2}} \simeq (U_B/W_B)/k_0 a$$

Weak absorption at the barrier means that the deflected angle can be large for the orbit for which the absorption probability is one-half, that is,

$$W_B/U_B \ll (k_0 a)^{-1} \tag{3.56}$$

Table 1.1 shows that $k_0 a \sim 1$ in most heavy-ion elastic scattering experiments, hence a simple condition for weak absorption at the barrier is $W_B \ll U_B$. A potential which is weakly absorbing at the barrier is often called a 'surface transparent potential'.

III. $V_B(l)$ smaller than $E$: For small angular momenta the barrier in the effective potential $V_{\text{eff}}$ is lower than the incident energy and the internal region $r < r_B$ of the optical potential can be explored by the collision. The behaviour of $S$ in this angular momentum range is influenced strongly by the imaginary part of the optical potential. In semi-classical theory it is possible to decompose $S$ into a part $S_B$ due to reflection at the barrier at $r_B$ and a part $S_I$ due to waves which penetrate into the internal region $r < r_B$, (Brink & Takigawa, 1977; Brink *et al.*, 1978),

$$S_l = S_{Bl} + S_{Il} \tag{3.57}$$

The magnitude of the barrier part $S_B$ depends on the surface diffuseness of the optical potential. The barrier contribution is enhanced if the surface is sharp, that is, if the surface diffuseness $a$ is small. The optical potential is approximately exponential in shape near the barrier and the magnitude of $S_B$ can be estimated from equation (3.39) for the reflection coefficient from an exponential potential. The result is

$$|S_{Bl}| \approx \exp(-2\alpha k_B a) \tag{3.58}$$

where $\alpha = \pi - \tan^{-1}(W_B/U_B)$, and $k_B$ is the local radial wave number at the barrier.

The magnitude of the internal part $S_I$ is very sensitive to the imaginary part of the optical potential. If $W(r)$ is large the internal wave is almost completely absorbed. As $2W/\hbar$ is the absorption probability per unit time and $2R_s/\bar{v}$ is an estimate of the time spent in the interaction region

$$|S_{Il}| \approx \exp(-2R_s\bar{W}/\hbar\bar{v}) \tag{3.59}$$

where $\bar{W}$ is an average strength of $W(r)$ and $\bar{v}$ is an average relative velocity in the interaction region. The strong interaction radius $R_s$(1.1) is an estimate of the radius of the interaction region. If the real potential U is deep then $\bar{v} \approx \bar{U}/\hbar k$ where $\hbar\bar{k} = \mu\bar{v}$ so that the estimate (3.59) can also be written as

$$|S_{Il}| \approx \exp(-\bar{k}R_s\bar{W}/\bar{U}) \tag{3.60}$$

In most cases $\bar{k}R_s$ is a large number. Hence, the internal amplitude $S_I$ is negligible unless $\bar{W} \ll \bar{U}$.

The potential curves for $l_1$, $l_2$, $l_3$ shown in fig. 3.6 have a barrier with a potential pocket behind it. The barrier disappears at a critical angular momentum $l_c$ and is absent if $l > l_c$. The critical energy $E_c$ is the energy of the highest barrier. Classical orbiting can occur if $E < E_c$ but not if $E > E_c$ (Miller 1969). Fig. 3.7(a) shows a typical classical deflection function for $E < E_c$. The singularity at $l = l_0$ corresponds to classical orbiting. If $l > l_0$ there is reflection at the outer barrier and when $l < l_0$ the relative motion penetrates into the pocket behind the barrier. The classical deflection angle $\Theta = 0$ when $l = 0$. Then the projectile passes straight through the target. When $E > E_c$ the deflection function has the form shown in fig. 3.7(b). The orbiting singularity has disappeared and is replaced by a dip in the deflection function. The nuclear rainbow effect, to be discussed in section 4.8, is associated with a deflection function like the one shown in fig. 3.7(*b*). Orbiting will be discussed in sections 6.3 and 6.4.

A strongly absorbing optical potential has the property that the imaginary part is comparable in magnitude with the real part in the barrier region. Estimates (3.56) and (3.60) show that for such a potential barrier penetration and orbiting effects need not be considered and that the

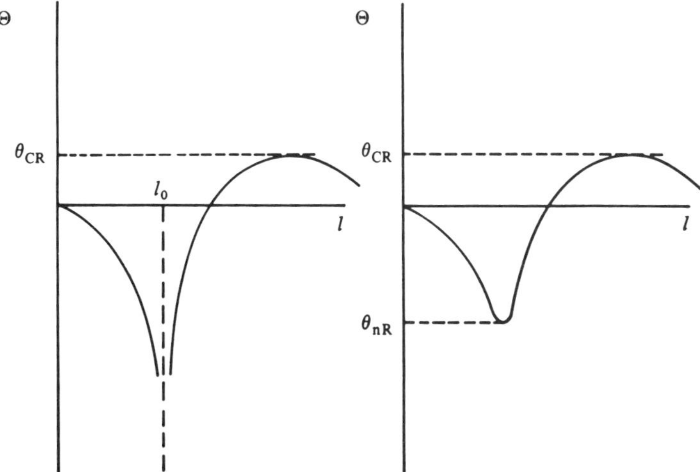


Fig. 3.7. Typical classical deflection functions. (*a*) $E < E_c$: the singularity at $l =$ corresponds to classical orbiting, (*b*) $E > E_c$: there is no orbiting but the minimum corresponds to a nuclear rainbow at $\theta_{nR}$. In both cases $\theta_{CR}$ is the Coulomb rainbow angle.

internal amplitude in (3.57) is negligible. The potential for ($^{12}C$, $^{208}Pb$) elastic scattering discussed in section 2.4 is a typical strong absorption potential and $|S_l|$ plotted in fig. 2.5 is a typical strong absorption amplitude. In the remainder of this section we derive some simple semi-classical formulae for phase shifts and partial wave amplitudes for a strongly absorbing potential. The analysis of surface transparent and weakly absorbing potentials is more complex and we postpone it to chapter 6.

We use formulae from section 3.8 to obtain semi-classical estimates for phase shifts. In the case of a strongly absorbing potential only the outermost turning point contributes and it is the one which should be used in equation (3.49). Other turning points give no contribution because of the large imaginary potential.

When the incident energy $E$ is well below the barrier $V_B(\lambda)$ the nuclear phase is small and can be calculated from (3.49) by a perturbation method. A modification $\Delta V$ in the nuclear potential gives a change $\Delta\delta$ in the WKB phase. To first order

$$\Delta\delta = -(\mu/\hbar^2)\int_{r_0}^{\infty}(\Delta V(r)/k(r))\,dr$$

If the integration variable is changed from $r$ to $t$ using $\hbar k(r) = \dot{r}$ then

$$\Delta\delta = -1/(2\hbar)\int_{-\infty}^{\infty}\Delta V(\mathbf{r}(t))\,dt \tag{3.61}$$

where the time integral is taken along a classical orbit $\mathbf{r}(t)$ determined by the unperturbed potential $V(r)$. If the whole nuclear potential $V_n(r)$ is taken as a perturbation the WKB phase is

$$\delta_n(\lambda) = -(\mu/\hbar^2)\int_{r_C}^{\infty} (V_n(r)/k_C(r))\,dr \tag{3.62}$$

$$= -1/(2\hbar)\int_{-\infty}^{\infty} V_n(r_C(t))\,dt \tag{3.63}$$

where the integral is taken along the Coulomb orbit $\mathbf{r}_C(t)$ with $\lambda = l + \frac{1}{2}$.

The imaginary part of a nuclear optical potential is absorptive ($\text{Im}\,V_n < 0$). Equation (3.62) shows that $\text{Im}\,\delta_n > 0$ so that the magnitude of the partial wave amplitude

$$|S(\lambda)| = \exp(-2\,\text{Im}\,\delta_n(\lambda)) \leqslant 1$$

Equation (3.62) can be used to estimate the nuclear phase when $V_n$ is small. Suppose that the nuclear potential has a Woods–Saxon form

$$V_n(r) = -(V_0 + iW_0)/(1 + \exp((r-R)/a))$$

and that we are interested in waves for which the Coulomb distance of closest approach $r_C$(2.2) is greater than $R$. Then the Woods–Saxon potential can be approximated by an exponential shape

$$V_n(r) \approx -(V_0 + iW_0)\exp((R - r/a) \tag{3.64}$$

and the formula (3.62) gives

$$2\delta_n = -(2\mu/\hbar^2)V_n(r_C)\int_{r_C}^{\infty} k_C^{-1}(r)\exp((r_C - r)/a)\,dr \tag{3.65}$$

If the surface diffuseness $a$ is small compared with $r_C$ then the main contribution to the integral comes from $r$ near $r_C$ and $k_C(r)$ (3.45) can be approximated by

$$k_C(r) \approx (k/r_C)\{2(r - r_C)(r_C - a_C)\}^{\frac{1}{2}}$$

Then the integral (3.65) can be calculated by the substitution $r - r_C = x^2$ to give

$$2\delta_n(\lambda) = -(1/\hbar v)\{2\pi a r_C/(1 - a_C/r_C)\}^{\frac{1}{2}}V_n(r_C) \tag{3.66}$$

In (3.66) $v = \hbar k/\mu$ is the asymptotic relative velocity. The factor $(1 - a_C/r_C) = (1 - V_C/2E)$ in the denominator takes into account the change in relative velocity and the curvature of the orbit of relative motion caused by the Coulomb interaction.

The length $r_C$ in (3.66) is the distance of closest approach in a Coulomb orbit with angular momentum $\lambda = l + \frac{1}{2}$

$$kr_C = n + (n^2 + \lambda^2)^{\frac{1}{2}} \tag{3.67}$$

A simple formula for the angular momentum dependence of $\delta_n$ can be obtained by expanding about a grazing orbit with angular momentum $\lambda_g$ and distance of closest approach $r_g$. From (3.67)

$$r_C - r_g \simeq (\lambda - \lambda_g)\lambda_g/\{k(n^2 + \lambda_g^2)^{\frac{1}{2}}\}$$

and (3.66) can be written as

$$2\delta_n(\lambda) = 2\delta_n(\lambda_g)\exp((\lambda_g - \lambda)/\Delta) \tag{3.68}$$

with

$$\Delta = ka(n^2 + \lambda_g^2)^{\frac{1}{2}}/\lambda_g \tag{3.69}$$

(We replace $r_C$ by $r_g$ in the factor in square brackets in (3.66) because it is a slowly varying function of $r_C$.) In section 5.3 we discuss a parametrization of the nuclear phase due to Kaufmann (1977). It is interesting because (3.69) gives a simple relation between $\Delta$ and the surface diffuseness of the underlying optical potential, and the phase of $\delta_n(\lambda)$ is given by

$$\arg\delta_n(\lambda) = \tan^{-1}(W_0/V_0) \tag{3.70}$$

The approximate formula (3.66) was obtained by integrating the nuclear potential along a Rutherford orbit. It does not take into account the effects due to modification of the orbit by the nuclear potential. The first correction in a power series expansion in the strength of the nuclear force was given by Brink & Satchler (1981) and can be obtained from (3.66) by the replacement

$$V_n(r_C) \to V_n(r_C)\left[1 - \frac{V_n(r_C)r_C}{2\sqrt{2E}\cdot a\cdot(1 - a_C/r_C)}\right] \tag{3.71}$$

where $E$ is the centre-of-mass energy. The correction (3.71) increases absorption because the attractive nuclear force pulls the projectile nearer the target where the imaginary potential $W(r)$ is stronger.

The semi-classical formula for the nuclear phase shift has been studied in a numerical example by Landowne *et al.* (1976). Fig. 3.8 shows $|S_{nl}| = \exp(-2\mathrm{Im}\,\delta_{nl})$ for the scattering of $^{18}$O and $^{58}$Ni at a bombarding energy of 60 MeV. The potential parameters are $V_0 = 90.1$ MeV, $W_0 = 42.9$ MeV, $R_v = R_w = 7.92$ fm and $a_v = a_w = 0.5$ fm. The solid dots result from integrating the Schrödinger equation numerically. The open circles are calculated from the WKB equation (3.49). The diagram shows that the WKB formula gives values which are very similar to the exact quantal results. Landowne *et al.* (1976) evaluate the semi-classical phase by integrating along the contour $C$ shown in fig. 3.9.

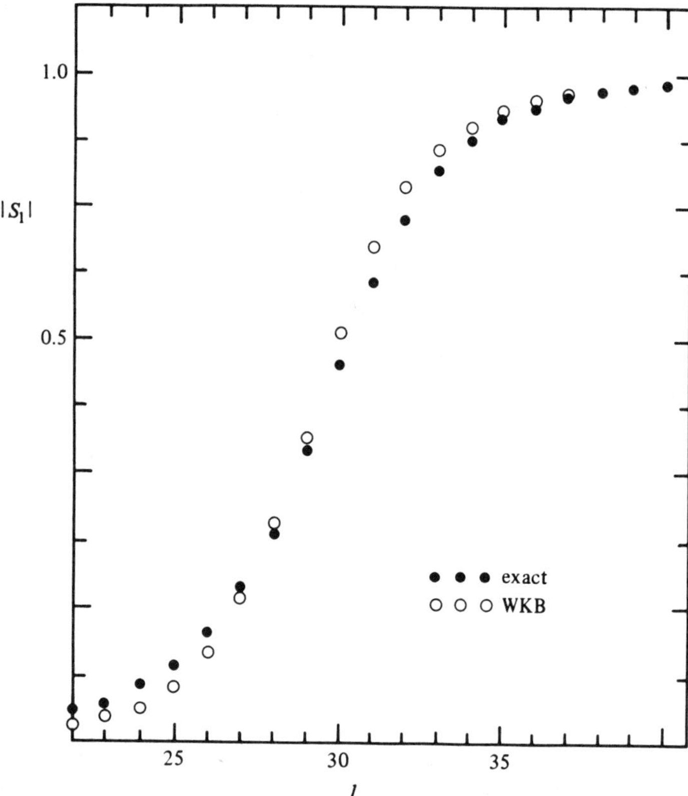


Fig 3.8. The magnitude of the nuclear reflection coefficient $|S_l|$ for $^{16}O + {}^{58}Ni$ at $E_{lab} = 60\,\text{MeV}$. The full dots are the result from integrating the Schrödinger equation numerically. The open dots are calculated in the WKB approximation using equation (3.49) (from Landowne *et al.*, 1976).

When the incident energy is near to or above the barrier $V_B(\lambda)$ the WKB nuclear phase is still given by equation (3.49), but the perturbation method cannot be used to calculate it. There is another approximation suggested by Anni *et al.* (1978) which is based on the idea that the scattered wave is reflected from the barrier region of the optical potential. Fig. 3.9 shows the real and imaginary parts of the complex turning point $r_0$ as a function of $\lambda$ for the optical potential for ($^{18}O$, $^{58}Ni$) scattering at $E_{lab} = 60\,\text{MeV}$ which was used in fig. 3.8. The real part of the turning point position $\text{Re}\,r_0$ breaks away from the Coulomb turning point $r_C$ for $\lambda < \lambda_g$ and tends to a limiting value of about 9 fm. The imaginary part $\text{Im}\,r_0$ tends to zero for $\lambda > \lambda_g$ and to the limiting value $-a\alpha$ predicted by (3.38) for $\lambda < \lambda_g$.

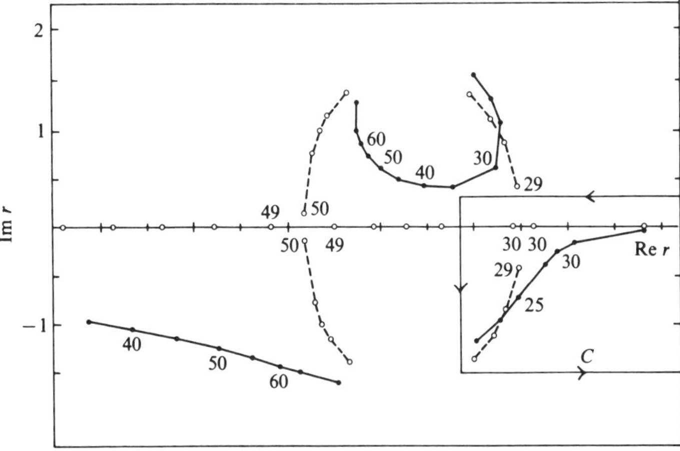


Fig. 3.9. Complex turning points for $^{16}O + ^{58}Ni$ at $E_{lab} = 60$ MeV using the same potential as for fig. 3.8. The open dots show the turning points when the imaginary potential is set equal to zero while the full dots are for the complex potential. The numbers give the angular momentum for various partial waves. The solid line $C$ with arrows shows part of the integration path used in the WKB calculations. It returns to the real axis outside the range of the interaction (From Landowne *et al.*, 1976).

Anni and co-workers approximate the WKB phase integral in (3.49) by writing

$$\int_{r_0}^{R} k(r)\,dr = \int_{r_0}^{R} (k(r) - k_C(r))\,dr + \int_{r_0}^{R} k_C(r)\,dr \tag{3.72}$$

The main contribution to the first integral on the right hand side of (3.72) comes from near the turning point $r_0$. It can be approximated by replacing the Coulomb and centrifugal potentials by their values at the strong absorption radius $r_g$. The resulting integral in the case of the exponential nuclear potentials (3.64) is the same as the one calculated in section 3.5. Its value is $k_g a(\ln 2 - 1) = -0.31 k_g a$ where

$$k_g \simeq k_C(r_g) = k(1 - V_C(r_g)/E)^{\frac{1}{2}} \simeq (\lambda_g^2 - \lambda^2)^{\frac{1}{2}}/r_g$$

The second integral gives a phase shift $2\sigma(\lambda, r_0)$ corresponding to scattering by a charged sphere of radius $r_0$. if $r_0$ is near $r_g$

$$2\sigma(\lambda, r_0) \simeq 2\sigma(\lambda, r_g) - 2k_g(r_0 - r_g)$$

The resulting estimate for the total phase contributed by the turning point $r_0$ is

$$2\delta(\lambda) = 2(\sigma(\lambda) + \delta_n(\lambda)) = 2\sigma(\lambda, r_g) - 2k_g(r_0 - r_g - 0.31\text{a}) \tag{3.73}$$

The first term in (3.73) is the phase shift for scattering by a charged hard sphere of radius $r_g$ and the second term is a correction due to the nuclear potential. The imaginary part

$$2\,\mathrm{Im}\,\delta = -2k_g\,\mathrm{Im}\,r_0 \tag{3.74}$$

reduces to (3.58) when $V_B(\lambda) \ll E$. The calculation of Anni *et al.* (1978) uses a Woods–Saxon nuclear potential rather than the simpler exponential form, but their result is qualitatively similar to (3.73.)

Equation (3.74) or the simpler (3.58) can be used to give an estimate of the reflection coefficient of the barrier when $V_B(\lambda) \ll E$,

$$|S(\lambda)| \simeq \exp(4k_g\,\mathrm{Im}\,r_0) \simeq \exp(-4k_g a\alpha) \tag{3.75}$$

# 4
# The semi-classical scattering amplitude

## 4.1 The structure of the amplitude

In a classical description of heavy-ion scattering a trajectory of relative motion of the two nuclei is obtained by solving the classical equations of motion under the influence of the Coulomb potential and the real part of the optical potential. The trajectory is specified by the centre-of-mass energy $E$ and the impact parameter $b$ or the angular momentum $\lambda = kb$. The scattering angle $\Theta(\lambda, E)$ is a function of $\lambda$ and $E$ and the classical cross-section is

$$\sigma_{\mathrm{cL}}(\theta) = d\sigma/d\Omega = (k^2 \sin\theta)^{-1}(\lambda d\lambda/d\Theta) \tag{4.1}$$

In the case of Coulomb scattering $\lambda$ and $\Theta$ are related by equation (2.2). Evaluating (4.1) yields the Rutherford cross-section (2.4).

Various reactions may take place which remove flux from the elastic channel. According to the interpretation in section 2.2 the probability $w$ per unit time for a transition out of the elastic channel is related to the imaginary part of the optical potential (2.6) by $w = 2W(r)/\hbar$. If $P(t)$ is the probability that the interacting nuclei remain in their ground states after a time $t$ then

$$\frac{dP}{dt} = -wP. \tag{4.2}$$

Integrating this equation gives

$$P(t) = \exp\left(-\int_{-\infty}^{t} w(t)\,dt\right) \tag{4.3}$$

Hence the probability that the two nuclei in a heavy-ion collision remain in their ground states after the scattering is $P_0 = P(\infty)$. This probability depends on the angular momentum $\lambda$ and using (4.3)

$$P_0(\lambda) = \exp\left(-(2/\hbar)\int_{-\infty}^{\infty} W(\mathbf{r}(\lambda, t))\,dt\right) \tag{4.4}$$

The integral (4.4) is taken along a classical orbit $\mathbf{r}(\lambda, t)$ in the real optical potential with angular momentum $\lambda$. Combining (4.1) and (4.4) gives a classical formula for the elastic cross-section

$$\sigma(\theta) = \sigma_{cL}(\theta) P_0(\lambda(\theta)) \tag{4.5}$$

where $\lambda(\theta)$ is the angular momentum of the orbit with scattering angle $\theta$.

We derive formulae for the semi-classical scattering amplitude in the present chapter and, in simple cases, show that

$$f(\theta) = (\sigma(\theta))^{\frac{1}{2}} \exp(i\alpha(\theta)) \tag{4.6}$$

where $\sigma(\theta)$ is given by (4.5) and $\alpha(\theta)$ is a phase which is characteristic of the classical trajectory with scattering angle $\theta$. The semi-classical elastic scattering cross-section

$$\sigma(\theta) = |f(\theta)|^2 \tag{4.7}$$

is identical with the classical result (4.5). Equation (4.6) is derived by using the semi-classical phase shifts (3.48) in the partial wave series for $f(\theta)$. The series is converted into a sum of integrals using the Poisson sum formula in section 4.2 and these integrals are evaluated by the stationary phase approximation in section 4.3.

The simple semi-classical formula (4.6) can be generalized to take into account the effects of interference and diffraction. If several classical trajectories with different angular momenta $\lambda_i$ have the same observation angle $\theta$ then the semi-classical theory developed in section 4.3 yields an amplitude

$$f(\theta) = \sum_i f_i(\theta) \tag{4.8}$$

where each of the terms $f_i(\theta)$ is given by a formula like (4.6) calculated from the classical trajectory with angular momentum $\lambda_i$, and the sum is taken over all trajectories with the same classical observation angle $\theta$. Rainbow scattering which is discussed in section 4.4 is an example of a case where two trajectories have the same scattering angle and the semi-classical amplitude (4.8) is a sum of two terms.

Rainbow scattering is also a case where the classical cross-section has a shadow region. For scattering angles $\theta$ less than a critical angle $\theta_R$ (the rainbow angle) two classical trajectories contribute to the cross-section. On the other hand, no classical orbits have a scattering angle $\theta > \theta_R$. There is a shadow for this angular region and the classical cross-section is zero. In quantum theory the cross-section is non-zero for $\theta > \theta_R$ because a diffracted wave spreads into the shadow region. The diffracted amplitude is estimated in section 4.4 by evaluating the semi-classical integral by the saddle point approximation. The saddle point occurs at a complex angular momentum and there is a sense in which the diffracted amplitude is associated with a 'complex classical trajectory' (section 4.5).

When deriving the semi-classical formula (4.6) and its generalization (4.8) the imaginary part of the optical potential is treated as a perturbation. It produces attenuation along a classical orbit determined by the real potential, but equation (4.6) does not take into account the effects of the imaginary potential on the orbit itself. This is done in section 4.5 by making more consistent use of the saddle point approximation and complex trajectories. It allows diffraction effects due to absorption to be accounted for in the semi-classical description.

The classical cross-section at the rainbow angle $\theta_R$ diverges and the semi-classical formula (4.8) breaks down completely for $\theta$ near $\theta_R$. A more general result which reduces to (4.8) for $\theta \ll \theta_R$ and is valid near the rainbow angle can be obtained by using a uniform approximation. The ideas behind the uniform approximation and the application to rainbow scattering are discussed in section 4.4.

Many of the semi-classical techniques used in this chapter were developed in the pioneering works of Mott & Massey (1949) and Ford & Wheeler (1959). The Poisson sum formula was first applied to heavy-ion reactions by Venter (1963). It was used by Frahn & Venter (1963) and later by many other authors. A number of the methods discussed in this chapter were developed and applied to heavy-ion reactions by the Copenhagen group (Broglia & Winther, 1972, 1974*a*, 1974*b*; Landowne *et al.*, 1976). da Silveira (1973) showed how to calculate the Coulomb rainbow cross-section in a simple way. The complex trajectory approaches have been used in diffraction theory for many years (Keller, 1958; Nussenzweig, 1979). They were applied to heavy-ion reactions by Knoll & Schaeffer (1976), Koeling & Malfliet (1975), Anni *et al.* (1974, 1976) and Rowley & Marty (1976*a*).

## 4.2 The Poisson sum formula

The initial step is to evaluate the scattering amplitude by finding approximate methods for summing the partial wave series

$$f(\theta) = 1/(2ik)\sum_l (2l+1)P_l(\cos\theta)(S_l - 1) \tag{4.9}$$

First, we give some exact results. Using the recurrence relation for Legendre polynomials

$$(2l+1)xP_l(x) = (l+1)P_{l+1}(x) + lP_{l-1}(x)$$

it is easy to show that

$$(1-\cos\theta)f(\theta) = 1/(2ik)\sum_l P_l(\cos\theta)[(2l+1)S_l - lS_{l-1} - (l+1)S_{l+1}] \tag{4.10}$$

This formula was used by Yennie *et al.* (1954) in order to accelerate the convergence of the partial wave series for electron scattering. It has also been applied in heavy-ion scattering (see Anni & Renna, 1981*a*, *b*). It gives an exact formula for the scattering amplitude in a sharp cut-off model

$$S_l = 0; \quad l < L, \quad S_l = 1; \quad l \geqslant L$$

Substituting this amplitude in (4.10) gives

$$f(\theta) = 1/(2ik)[P_L(\cos\theta) - P_{L-1}(\cos\theta)]/(1 - \cos\theta)$$

Another application is to show that

$$\sum_l (2l+1) P_l(\cos\theta) = 2\delta(1 - \cos\theta) \tag{4.11}$$

Equation (4.10) shows that the sum in (4.11) is zero unless $\cos\theta = 1$. Integrating both sides of (4.11) with respect to $\cos\theta$ between the limits $-1$ and 1 checks the normalization. An alternative way to obtain the series (4.11) is to use the result

$$\sum_l (2l+1) h^l P_l(\cos\theta) = (1 - h^2)(1 - 2h\cos\theta + h^2)^{-\frac{3}{2}}$$

When $h$ is near unity the right-hand side of this equation is a very sharply peaked function around $\theta = 0$, with height $(1+h)/(1-h)^2$ and width $\Delta\theta \simeq 1 - h$. As $h \to 1$ it tends to a $\delta$-function as in (4.11). Equation (4.11) has the consequence that the term $-1$ in the factor $S_l - 1$ can be omitted in the formula (4.9) for the scattering amplitude provided $f(\theta)$ is evaluated away from the forward direction.

The Poisson sum formula has been used by Venter (1963), by Frahn (1984), by Rowley & Marty (1976*a*) and by other authors as a starting point for approximate evaluations of the partial wave sum. Suppose that $a(\lambda)$ is a continuous function of $\lambda$ for $0 < \lambda < \infty$ and $a_l = a(l + \frac{1}{2})$. The Poisson sum formula states

$$\sum_{l=0}^{\infty} a_l = \sum_{m=-\infty}^{\infty} (-1)^m \int_0^{\infty} d\lambda\, a(\lambda) \exp(2m\pi i\lambda) \tag{4.12}$$

It can be derived by writing $a(l + x)$ as a Fourier series for $0 < x < 1$

$$a(l + x) = \sum_{m=-\infty}^{\infty} A_m \exp(-2m\pi i x) \tag{4.13}$$

where the Fourier coefficients are given by

$$A_m = \int_l^{l+1} d\lambda\, a(\lambda) \exp(2m\pi i\lambda)$$

Putting $x = \frac{1}{2}$ in (4.13) and substituting the formula for $A_m$ gives

$$a_l = a(l + \tfrac{1}{2}) = \sum_{m=-\infty}^{\infty} (-1)^m \int_l^{l+1} d\lambda a(\lambda) \exp(2m\pi i\lambda)$$

Now, summing over $l$ gives the Poisson formula (4.12). If the function $a(\lambda)$ is analytic on and near the real axis for $\lambda > 0$ the formula can also be derived by using the Sommerfeld–Watson transformation (see Rowley & Marty, 1976*a*).

Many authors apply the Poisson formula directly to the partial wave sum (4.9). We prefer first to replace the Legendre polynomials in the sum by their asymptotic formulae which are valid for large $l$

$$P_l(\cos\theta) \sim \left[\frac{2}{\pi(l+\frac{1}{2})\sin\theta}\right]^{\frac{1}{2}} \cos((l+\tfrac{1}{2})\theta - \tfrac{1}{4}\pi); \quad 1/l \lesssim \theta \lesssim \pi - 1/l \tag{4.14}$$

By making this replacement before using the Poisson formula one avoids having to study the properties of Legendre polynomials with non-integer values of $l$. The Poisson formula gives the result

$$f(\theta) = (2/ik)(2\pi\sin\theta)^{-\frac{1}{2}} \sum_{m=-\infty}^{\infty} (-1)^m \times \int_0^\infty \lambda^{\frac{1}{2}} d\lambda \cos(\lambda\theta - \tfrac{1}{4}\pi) \exp(2m\pi i\lambda) S(\lambda) \tag{4.15}$$

In (4.15) $S(\lambda)$ is a continuous function of $\lambda$ with the property that $S(l + \frac{1}{2}) = S_l$ for integer values of $l$. The quantal scattering amplitudes $S_l$ are defined only for integer values of $l$, while (4.15) requires $S(\lambda)$ for all real positive values of $\lambda$. The interpolation required is supplied automatically by the semi-classical formulae (3.47) and (3.49) for the Coulomb and nuclear phases. In the final step of the present argument we write the cosine function in (4.15) in terms of exponential functions and obtain the formula

$$f(\theta) = \sum_{m=-\infty}^{\infty} (f_m^-(\theta) + f_m^+(\theta)) \tag{4.16}$$

where

$$f_m^{\pm}(\theta) = (-1)^m (1/ik)(2\pi\sin\theta)^{-\frac{1}{2}} \times \int_0^\infty \lambda^{\frac{1}{2}} d\lambda S(\lambda) \exp i(2m\pi\lambda \pm \lambda\theta \mp \tfrac{1}{4}\pi) \tag{4.17}$$

The representation (4.16) for the scattering amplitude is useful in semi-classical situations because normally the dominant contribution to $f(\theta)$ comes from only one or two terms in the series.

The asymptotic formula (4.14) is not valid for $\theta$ near 0 or $\pi$. The asymptotic formula in terms of the Bessel function $J_0$

$$P_l(\cos\theta) \sim (\theta/\sin\theta)^{\frac{1}{2}} J_0((l+\tfrac{1}{2})\theta) \tag{4.18}$$

is a good approximation if $\theta \lesssim \pi - 1/l$ and if $l$ is large. If $\theta$ is near $\pi$ then

$$P_l(\cos\theta) \sim (-1)^l[(\pi-\theta)/\sin\theta]^{\frac{1}{2}} J_0((l+\tfrac{1}{2})(\pi-\theta)) \tag{4.19}$$

## 4.3 The stationary phase approximation

Sometimes the integrals (4.17) can be evaluated by the method of stationary phase. The first discussion will be restricted to the case where $\delta(\lambda)$ is real and we write equation (4.17) in the form

$$f_m^{\pm}(\theta) = \int_0^{\infty} g(\lambda)\exp(i\psi(\lambda))\,d\lambda \tag{4.20}$$

$$\psi(\lambda) = 2\delta(\lambda) \pm \lambda\theta + 2m\pi\lambda \mp \tfrac{1}{4}\pi - (m+\tfrac{1}{2})\pi \tag{4.21}$$

$$g(\lambda) = [\lambda/(2\pi k^2 \sin\theta)]^{\frac{1}{2}} \tag{4.22}$$

The more general case will be considered later. The stationary phase approximation is obtained by assuming that the main contributions to the integral (4.20) come from parts of the range of integration near points where the phase $\psi(\lambda)$ is stationary, because there the contributions from different values of $\lambda$ will be nearly in phase and will add constructively. In other parts of the range of integration where $\psi(\lambda)$ changes rapidly there will be destructive interference between contributions from nearby values of $\lambda$ and those parts will give a small contribution to the integrals. The steps in the stationary phase approximation are (Appendix B) to

(a) locate the stationary points $\lambda_s$ by solving

$$\psi'(\lambda_s) = (d\psi/d\lambda)_{\lambda_s} = 0, \tag{4.23}$$

(b) expand $\psi(\lambda)$ as a Taylor series about each stationary point $\lambda_s$ up to quadratic terms,

(c) assume that $g(\lambda)$ is slowly varying so that it can be replaced by $g(\lambda_s)$ when calculating the contribution of the stationary point $\lambda_s$.

The resulting approximate formula for the integral (4.20) is

$$f_m^{\pm}(\theta) = \sum_s g(\lambda_s)(2\pi/|\psi''(\lambda_s)|)^{\frac{1}{2}} \exp(i\psi(\lambda_s) - i\chi_s) \tag{4.24}$$

$$\chi_s = -\tfrac{1}{4}\pi \qquad \text{if} \qquad \psi''(\lambda_s) > 0$$

$$\chi_s = \tfrac{1}{4}\pi \qquad \text{if} \qquad \psi''(\lambda_s) < 0$$

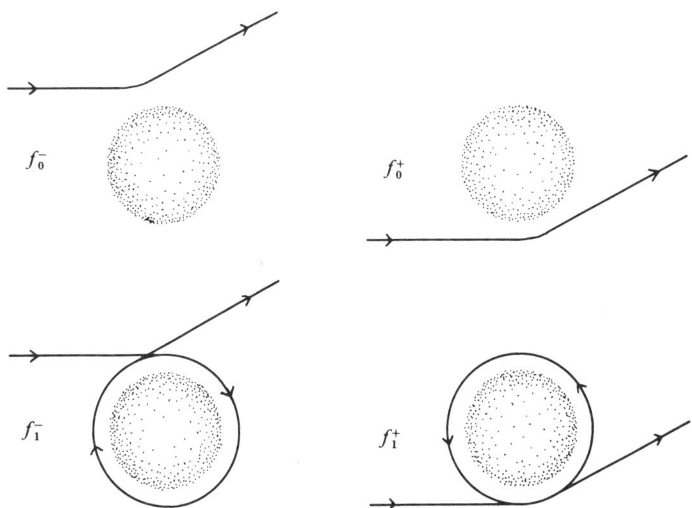


Fig. 4.1. Illustration of the classical orbits corresponding to several terms in the Poisson sum formula for the scattering amplitude $f(\theta)$. The amplitudes $f_0^-$ and $f_1^-$ are nearside while $f_0^+$ and $f_1^+$ are farside.

The sum in (4.24) is taken over all solutions of the stationary condition (4.23) in the range $0 < \lambda_s < \infty$. Using equation (4.21) the stationary phase condition (4.23) is

$$2\delta'(\lambda_s) = \mp\,\theta - 2m\pi \tag{4.25}$$

This condition is identical to (3.54) which relates the semi-classical phase to the classical scattering angle. Hence, the classical paths with scattering angle corresponding to the angle of observation $\theta$ give the main contribution to $f(\theta)$ in the stationary phase limit. A stationary point in $f_0^-(\theta)$ corresponds to scattering through an angle $\theta$ while $f_0^+(\theta)$ corresponds to scattering through $-\theta$. Stationary points in $f_m^\pm(\theta)$ correspond to classical orbits which circulate the origin $m$ times. This gives a physical picture of other terms in the Poisson series and helps to decide when it is necessary to include them. Fig. 4.1 illustrates this interpretation of several terms in the Poisson series.

The amplitudes $f_m^-(\theta)$ are often called nearside amplitudes while $f_m^+(\theta)$ are farside amplitudes. There are other notations. Fuller (1975), Hussein & McVoy (1984) and others use

$$f_{\mathrm{N}} \equiv f^- \quad \text{and} \quad f_{\mathrm{F}} \equiv f^+ \tag{4.26}$$

for the nearside and farside amplitudes. Frahn (1984) uses the opposite convention ($f^+$ for nearside, $f^-$ for farside).

We obtain the stationary phase formula for $f(\theta)$ by using the expressions (4.21) and (4.22) for $g(\lambda)$ and $\psi(\lambda)$ in (4.24) and then summing all the Poisson components in (4.16). The pre-exponential factor in (4.24) is

$$\left[\frac{\lambda_s}{2k^2 \sin\theta |\delta''(\lambda_s)|}\right]^{\frac{1}{2}} = \left|\frac{\lambda_s}{k^2 \sin\theta} \cdot \frac{d\lambda_s}{d\theta}\right|^{\frac{1}{2}} \\ = |\sigma_{\text{cLs}}(\theta)|^{\frac{1}{2}} \tag{4.27}$$

where $\sigma_{\text{cL}}$ is the classical cross-section (4.1). The phase is

$$\alpha_s(\theta) = 2\delta(\lambda_s) \pm \theta\lambda_s + 2m\pi\lambda_s - \chi_m^{\pm} \tag{4.28}$$

$$\left.\begin{aligned} \chi_m^- &= m\pi, & \chi_m^+ &= (m+\tfrac{1}{2})\pi \quad \text{if} \quad \delta'' > 0 \\ \chi_m^- &= (m+\tfrac{1}{2})\pi, & \chi_m^+ &= (m+1)\pi \quad \text{if} \quad \delta'' < 0 \end{aligned}\right\} \tag{4.29}$$

Substituting in (4.16) gives

$$f(\theta) = \sum_s (\sigma_{\text{cLs}})^{\frac{1}{2}} \exp(i\alpha_s(\theta)) \tag{4.30}$$

which is the form anticipated in equations (4.7) and (4.8) of section 4.1. The sum in (4.30) is taken over all $\lambda_s$ which are stationary points of any one of the integrals (4.17). In physical terms the sum is over all classical trajectories with observation angle $\theta$.

A semi-classical formula for the Coulomb scattering amplitude can be obtained by using these results. The semi-classical Coulomb phase (equation (3.47)) is

$$\sigma(\lambda) = \tfrac{1}{2}n \ln(n^2 + \lambda^2) - n + \lambda \tan^{-1}(n/\lambda)$$

and the stationary point of the integral for the nearside amplitude $f_0^-(\theta)$ is determined by equation (4.25)

$$2\sigma'(\lambda_s) = 2\tan^{-1}(n/\lambda_s) = \theta \tag{4.31}$$

Equation (4.31) is the classical relation (2.2) between angular momentum and scattering angle for Coulomb scattering. In this example

$$\sigma''(\lambda_s) = -n/(n^2 + \lambda_s^2) = -(1/n)\sin^2\tfrac{1}{2}\theta \tag{4.32}$$

so that $\chi_0^- = \frac{1}{2}\pi$ and the phase

$$\begin{aligned} \alpha(\theta) &= 2\sigma(\lambda_s) - \theta\lambda_s - \tfrac{1}{2}\pi \\ &= n\ln(n^2 + \lambda_s^2) - 2n - \tfrac{1}{2}\pi \\ &= 2(n\ln n - n) - \tfrac{1}{2}\pi - n\ln\sin^2\tfrac{1}{2}\theta \end{aligned}$$

Equation (4.30) contains just one term in the summation and reduces to expression (2.5) for the Coulomb amplitude $f_0(\theta)$ with the Coulomb phase $\sigma_0(n)$ replaced by its asymptotic form

$$\sigma_0 \sim n\ln n - n + \tfrac{1}{4}\pi$$

which is valid for $n \gg 1$.

Next we consider the case where the nuclear phase is complex. The simplest approximation is obtained by assuming that $\exp(-2\,\mathrm{Im}\,\delta(\lambda))$ is a slowly varying function of $\lambda$ and including it in the pre-exponential factor $g(\lambda)$ in equation (4.20) so that (4.21) and (4.22) are replaced by

$$\left.\begin{aligned}\psi(\lambda) &= 2\,\mathrm{Re}\,\delta(\lambda) \pm \lambda\theta + 2m\pi\lambda \mp \tfrac{1}{4}\pi - (m+\tfrac{1}{2})\pi \\ g(\lambda) &= [\lambda/(2\pi k^2 \sin\theta)]^{\frac{1}{2}} \exp(-2\,\mathrm{Im}\,\delta(\lambda))\end{aligned}\right\} \tag{4.33}$$

The stationary phase evaluation of the integrals (4.20) proceeds just as before. Now the stationary values of $\lambda$ satisfy

$$2\,\mathrm{Re}\,\delta'(\lambda_s) = \mp\theta - 2m\pi$$

instead of (4.25). Formula (4.30) for the scattering amplitude is replaced by

$$\left.\begin{aligned}f(\theta) &= \sum_s (\sigma_s(\theta))^{\frac{1}{2}} \exp(i\alpha_s(\theta)) \\ \sigma_s(\theta) &= \sigma_{\mathrm{cL}s}(\lambda_s) P(\lambda_s)\end{aligned}\right\} \tag{4.34}$$

where $\sigma_{\mathrm{cL}}(\lambda)$ is given by (4.1) and $\alpha_s(\theta)$ by (4.28) with $\delta(\lambda_s)$ replaced by $\mathrm{Re}\,\delta(\lambda_s)$. The quantity

$$P(\lambda_s) = \exp(-4\,\mathrm{Im}\,\delta(\lambda_s))$$

and is the probability that there is no absorption along the classical path with angular momentum $\lambda_s$. If $\mathrm{Im}\,\delta(\lambda)$ is obtained from the perturbation formula (3.61) then $P(\lambda_s)$ is given by equation (4.4) and we obtain the results (4.6) and (4.8) in section 4.1.

The stationary values of the angular momentum given by equation (4.25) depend on the scattering angle $\theta$ and the centre-of-mass energy $E$. The phase $\alpha(\theta)$ in equation (4.28) is a function of $\theta$ and $E$. We conclude this section by giving two simple formulae for partial derivatives of the phase

$$\left(\frac{\partial\alpha}{\partial\theta}\right)_E = \pm\lambda_s, \quad \left(\frac{\partial\alpha}{\partial E}\right)_\theta = 2\left(\frac{\partial\delta}{\partial E}\right)_{\lambda_s} = \tau/\hbar \tag{4.35}$$

where $\tau$ is a time related to the collision time. The first formula is obtained by differentiating equation (4.28) remembering that $\lambda_s$ is a function of $E$ and

$$\left(\frac{\partial\alpha}{\partial\theta}\right)_E = \pm\lambda_s + \left\{2\left(\frac{\partial\delta}{\partial\lambda_s}\right)_E \pm\theta + 2m\pi\right\}\left(\frac{\partial\lambda_s}{\partial\theta}\right)_E$$

The coefficient of $(\partial\lambda_s/\partial\theta)_E$ is zero because of the stationary condition (4.25). The second formula follows in a similar way.

$$\left(\frac{\partial\alpha}{\partial E}\right)_\theta = 2\left(\frac{\partial\delta}{\partial E}\right)_{\lambda_s} + \left\{2\left(\frac{\partial\delta}{\partial\lambda_s}\right)_E \pm\theta + 2m\pi\right\}\left(\frac{\partial\lambda_s}{\partial E}\right)_\theta$$

Again the second term is zero because of (4.25). The arguments leading to equation (3.51) in section 3.8 show that the classical transit time $T(R)$ from a radial separation $R$ before the collision to a separation $R$ after the collision is

$$\begin{aligned} T(R) &= 2(\partial\delta/\partial E)\lambda_s + (kR + n\ln(2kR) + n)2E \\ &= \tau + T_0(R) \end{aligned} \tag{4.36}$$

The second term $T_0(R)$ depends only on $n$ and $E$ and not on the angular momentum or scattering angle. Hence, $\tau$ is the difference due to Coulomb and nuclear effects between $T(R)$ and the 'standard time' $T_0(R)$. The difference in collision time between two trajectories with the same energy is $\tau_1 - \tau_2$ because $T_0(R)$ cancels. The formulae (4.35) can be used to extract information about angular momenta and collision time from experimental data.

## 4.4 Rainbow scattering

The terms in the Poisson sum formula for the scattering amplitude $f(\theta)$ were evaluated by the stationary phase approximation in section 4.3. Stationary values $\lambda_s(\theta)$ of the angular momentum correspond to classical trajectories with scattering angle $\theta$. If there are several trajectories with the same scattering angle the stationary phase amplitude is a sum of contributions (4.34) from each of the trajectories

$$f(\theta) = \sum_s (\sigma_s(\theta))^{\frac{1}{2}} \exp(i\alpha_s(\theta)) \tag{4.37}$$

In equation (4.37) $\sigma_s(\theta)$ is the classical cross-section (4.5) due to the orbit $s$ modified by absorption, and $\alpha_s(\theta)$ is the semi classical phase (4.28). The cross-section

$$\sigma(\theta) = \sum_s \sigma_s(\theta) + 2\sum_{s<t} (\sigma_s\sigma_t)^{\frac{1}{2}} \cos(\alpha_s(\theta) - \alpha_t(\theta)) \tag{4.38}$$

calculated from (4.37) is a sum of contributions from each of the allowed classical trajectories together with interference terms.

Rainbow scattering occurs if the classical scattering angle as a function of angular momentum has the form shown in fig. 4.2. When $\lambda$ is large the deflection function is almost the same as the one for Coulomb scattering; but for smaller $\lambda$ it is modified by the attractive nuclear force and there is a maximum scattering angle $\theta_R$. For $\theta < \theta_R$ two angular momenta, $\lambda_a$ and $\lambda_b$, contribute at each scattering angle. The classical cross-section is shown in fig. 4.5. When $\theta$ is small $\lambda_b$ gives the larger contribution and the cross-section is almost equal to the Rutherford value. The classical cross-section has a singularity at $\theta_R$ because many angular momenta are focused to almost the same scattering angle. In mathematical terms the singularity is due to the vanishing of the derivative $(d\theta/d\lambda)$ in the formula (4.1) for the classical cross-section at the rainbow angle.

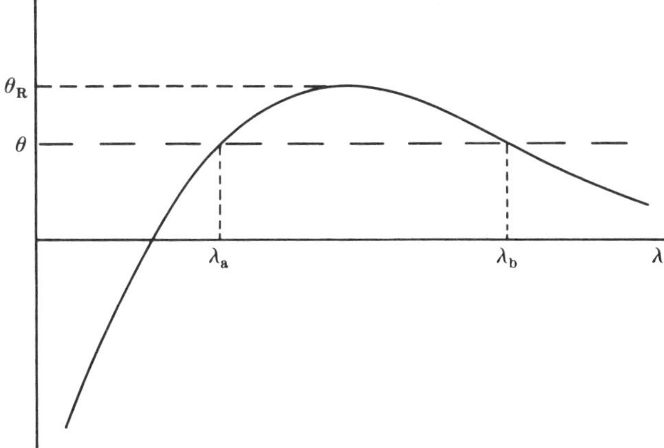


Fig. 4.2. Deflection function for rainbow scattering. $\theta_R$ is the rainbow angle. For $\theta < \theta_R$ two angular momenta $\lambda_a$ and $\lambda_b$ contribute at each scattering angle.

When $\theta < \theta_R$ the angular momenta $\lambda_a$ and $\lambda_b$ indicated in fig. 4.2 are stationary values of the phase $\psi(\lambda)$ in equation (4.20) for the nearside amplitude $f_0^-(\theta)$ and the stationary phase method gives an amplitude with two components

$$f(\theta) \approx f_0^-(\theta) \approx f_a^-(\theta) + f_b^-(\theta) \tag{4.39}$$

These interfere to give an oscillatory cross-section as indicated in equation (4.38). The cross-section including the interference term is shown by the dashed curve in fig. 4.5.

When $\theta > \theta_R$ the integral (4.20) has no stationary points, there are no classical trajectories with scattering angle $\theta$ and the classical cross-section is zero. In the semi-classical theory there is a diffractive amplitude $f_d(\theta)$ which spreads the cross-section into the classically forbidden region. This amplitude can be estimated by the saddle point method. Saddle points are stationary points of $\psi(\lambda)$ as a function of the complex variable $\lambda$. They are also stationary points of

$$|\exp i\psi(\lambda)| = \exp(-\operatorname{Im}\psi(\lambda)) \tag{4.40}$$

Fig. 4.3 is a sketch of the topography of the function (4.40) for two cases $\theta < \theta_R$ and $\theta > \theta_R$. When $\theta < \theta_R$ there are two saddle points on the real $\lambda$-axis which coincide with the stationary phase points $\lambda_a$ and $\lambda_b$ in fig. 4.2. The integration path in equation (4.20) is the real $\lambda$-axis between $\lambda = 0$ and $\lambda = \infty$. When evaluating the integral by the saddle point method the integration path is deformed in the complex $\lambda$-plane to the steepest descent path indicated by the thick line in fig. 4.3(*a*). The main contribution to

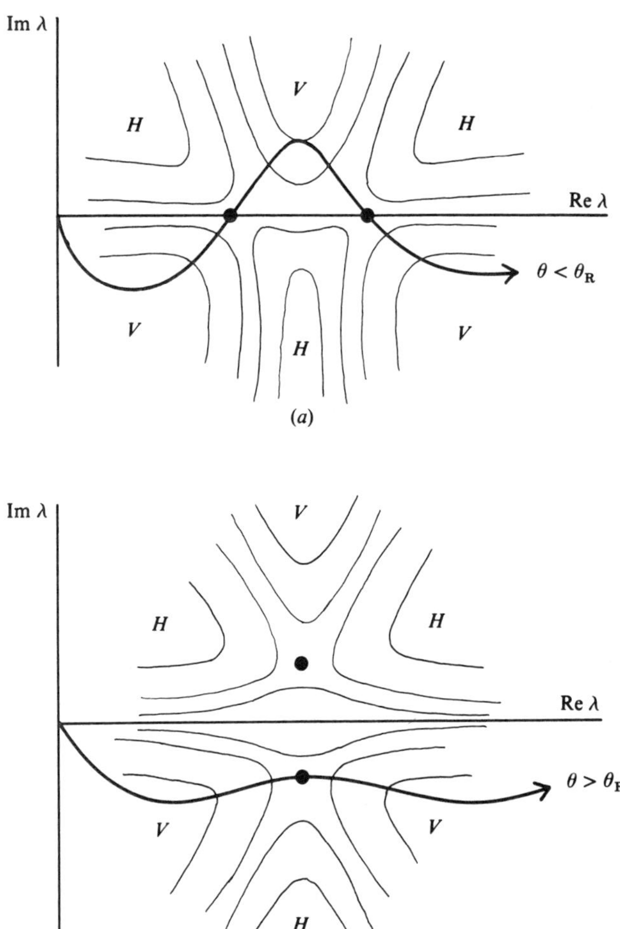


Fig. 4.3. The topography of the function $|\exp(i\psi(\lambda))|$ for complex $\lambda$ showing contours of constant altitude. The hills are indicated by $H$ and the valleys by $V$. The dots show the saddle point positions. The curve with an arrow shows the integration contour. It remains in the valleys except when crossing a saddle. (*a*) $\theta < \theta_R$: the saddle points are real, (*b*) $\theta > \theta_R$: the saddle points are complex.

the integral (4.20) comes from near the stationary values $\lambda_a$ and $\lambda_b$ in both the stationary phase and the saddle point methods, and both methods give the same approximate formula (4.24) (see appendix B). When $\theta > \theta_R$ there are two saddle points in the complex $\lambda$-plane indicated by black dots in fig. 4.3(*b*). Only one contributes to the integral, because the steepest descent path crosses only the saddle with $\text{Im}\,\lambda < 0$. When $\theta \approx \theta_R$ the two saddle points are close together, the condition for the validity of the saddle

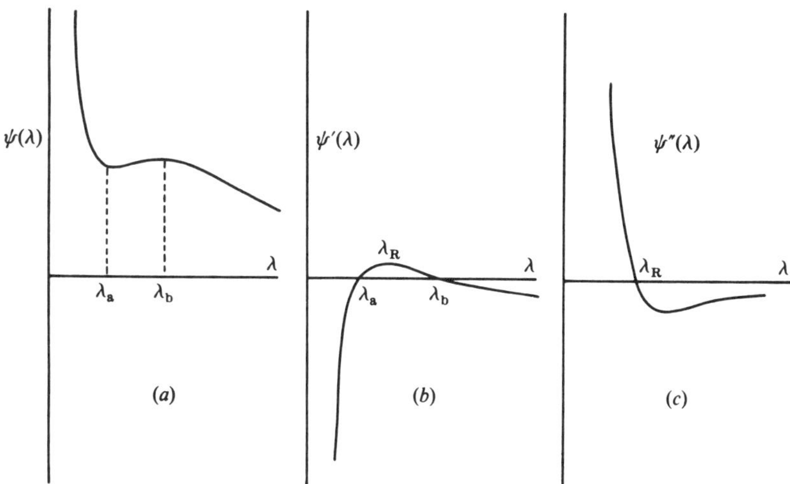


Fig. 4.4. Sketch of (*a*) the phase $\psi(\lambda)$ (equation 4.41) as a function of $\lambda$ and its derivatives, (*b*) $\psi'(\lambda)$, and (*c*) $\psi''(\lambda)$. The points of stationary phase ($\psi'(\lambda)=0$) are $\lambda_a$ and $\lambda_b$. The rainbow angular momentum ($\psi''(\lambda)=0$) is at $\lambda=\lambda_R$.

point method is not satisfied and it cannot be used. An expression of much wider validity can be obtained by means of the 'uniform approximation' developed by Chester *et al.* (1957) and by Berry (1966). This method will be presented now.

We want to calculate the nearside amplitude $f_0^-(\theta)$ so that the phase $\psi(\lambda)$ is

$$\psi(\lambda)=2\delta(\lambda)-\lambda\theta-\tfrac{3}{4}\pi$$

Figure 4.4 shows a sketch of a typical form for $\psi(\lambda)$ and its first two derivatives as a function of $\lambda$ when $\theta<\theta_R$. The stationary values of $\psi(\lambda)$ are zeros of $\psi'(\lambda)$, and $\psi(\lambda)$, and $\psi$ has a minimum $\psi_a$ at the stationary point $\lambda_a$ and a maximum $\psi_b$ at $\lambda_b$. The first derivative $\psi'(\lambda)$ has a maximum at the rainbow angular momentum $\lambda_R$ and $\psi''(\lambda)$ changes sign from positive to negative at $\lambda_R$. The Airy uniform approximation which is explained in appendix C gives

$$f_0^-(\theta)\approx\sqrt{\pi}[u\mathrm{Ai}(-\beta)-iv\mathrm{A'i}(-\beta)]\exp\{\tfrac{1}{2}i(\psi_a+\psi_b)\} \tag{4.41}$$

where Ai is an Airy function, A'i is its derivative and

$$\beta=[\tfrac{3}{4}(\psi_b-\psi_a)]^{\frac{2}{3}}$$
$$u=\beta^{\frac{1}{4}}[\sqrt{\sigma_a}+\sqrt{\sigma_b}]$$
$$v=\beta^{-\frac{1}{4}}[\sqrt{\sigma_a}-\sqrt{\sigma_b}]$$
$$\sigma_a=2\pi g^2(\lambda_a)/\psi''(\lambda_a),$$
$$\sigma_b=-2\pi g^2(\lambda_b)/\psi''(\lambda_b)$$

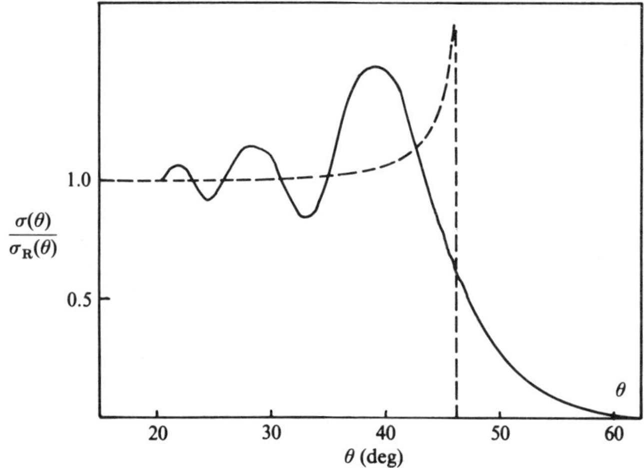


Fig. 4.5. A rainbow cross-section for Ar + S at $E_{lab} = 210$ MeV. The rainbow angle is at $\theta_R = 47°$. The dashed curve shows the classical rainbow cross-section.

In these expressions $\sigma_a$ and $\sigma_b$ are the classical cross-sections (4.1) (or (4.5) if $\mathrm{Im}\,\delta \neq 0$) evaluated at the stationary angular momenta $\lambda_a$ and $\lambda_b$. All the quantities appearing in equation (4.41) are functions of the scattering angle $\theta$. When $\theta < \theta_R$ $\beta, \sigma_a, \sigma_b$ are all positive while $u$ and $v$ are real. It can be shown that $\beta$, $u$ and $v$ are smooth functions of $\theta$ and that they remain real on the shadow side of the rainbow ($\theta > \theta_R$), even though the stationary points $\lambda_a$ and $\lambda_b$ become complex ($\lambda_b = \lambda_a^*$). On the shadow side of the rainbow $\beta < 0$. At $\theta = \theta_R$ the classical cross-section diverges but $u$ and $v$ remain finite. When $\theta$ is much less than $\theta_R$ (i.e. when $\beta \gg 1$) the Airy functions can be replaced by their asymptotic forms (A.7) and equation (4.41) reduces to the stationary phase formula,

$$f_0^-(\theta) \approx \sqrt{\sigma_a} \exp\{i\psi_a + \tfrac{1}{4}i\pi\} + \sqrt{\sigma_b} \exp\{i\psi_b - \tfrac{1}{4}i\pi\} \qquad (4.42)$$

The Airy uniform approximation (4.41) gives a smooth interpolation between the bright and shadow sides of the rainbow. It is interesting to notice that the uniform formula can be evaluated if $\psi_a, \psi_b, \sigma_a, \sigma_b$ are known. This is the same information which is needed for calculating the simple stationary phase approximation. The expression in which the information is used is more complicated, but the Airy functions are not difficult to obtain, either from tables or from a series expansion in powers of $\beta$.

The dashed curve in fig. 4.5 shows a typical rainbow angular distribution calculated from the Airy uniform approximation (da Silveira, 1973). Many experimental heavy-ion elastic distributions have this shape and are often called rainbow scattering distributions. Another interpretation due to Blair (1954) and Frahn (1966) describes them as Fresnel diffraction angular distributions. We discuss the Fresnel theory and its relation to the rainbow description in chapter 5.

The rainbow pattern in fig. 4.5 is often called a 'Coulomb rainbow' because it is associated with a modification of Coulomb scattering. The maximum in the deflection function in fig. 4.2 is caused by a strong Coulomb repulsion for large separations modified by a strong nuclear attraction for smaller separations. Another rainbow effect was proposed by Goldberg & Smith (1974) to explain the angular distributions for medium energy $\alpha$-nucleus scattering. It is called a nuclear rainbow and will be discussed in section 4.8.

## 4.5 Complex trajectories

The stationary phase evaluation (4.34) of the integrals (4.20) is a good approximation only if $g(\lambda)$ is a slowly varying function of $\lambda$. It can be inaccurate for a strongly absorbing potential because $\mathrm{Re}\,\delta(\lambda)$ and $\mathrm{Im}\,\delta(\lambda)$ vary equally rapidly with $\lambda$ and have comparable magnitudes. A more accurate result can be obtained by including both the real and the imaginary parts of the phase in $\psi(\lambda)$ and evaluating the resulting integral by the saddle point method. The saddle point positions are the solutions of equation (4.25), but now they are complex even when $\theta < \theta_R$ and $\lambda_a^* \neq \lambda_b$ when $\theta > \theta_R$. Indeed, even the rainbow angle $\theta_R$ is not defined precisely. The integration path should be deformed to pass through the saddle points as indicated in fig. 4.3. The complex saddle point positions can be associated with complex classical trajectories in the optical potential. The incident momentum and the scattering angle are real, but the classical orbit is complex and has a complex impact parameter. In this way the quantal diffraction effects due to absorption can be included in the semi-classical description. The complex trajectories can be regarded as 'fuzzy rays' representing diffractive wave-spreading (McVoy, 1978). They do not have to be calculated explicitly when the theory is applied as the input for calculations is the phase $\delta(\lambda)$. The saddle point evaluation of (4.20) can be made using the formula in appendix B with a result that is almost the same as the stationary phase formula (4.30)

$$f(\theta) = \sum_s \sqrt{\sigma_{\mathrm{cL}}(\lambda_s)} \exp\{i(\psi(\lambda_s) + \tfrac{1}{4}\pi)\} \tag{4.43}$$

In equation (4.43) $\sigma_{\mathrm{cL}}$ is given by the same formula

$$\sigma_{\mathrm{cL}}(\lambda_s) = (1/k^2 \sin\theta)(\lambda_s d\lambda_s/d\theta)$$

as a classical cross-section, but is complex because the stationary point $\lambda_s$ is complex. The sum in the stationary phase formula (4.30) is taken over all stationary points on the real $\lambda$-axis. The situation is more complicated with the saddle point method. The sum in (4.43) is over saddle points $\lambda_s$ of the integrals (4.20), but there can be many saddle points in the complex $\lambda$-plane and it is a problem to decide which ones should be included in the sum over $s$ in (4.43). Knoll and Schaeffer (1976) have made a careful study of this question and give a procedure for deciding which saddle points contribute. In cases where the stationary phase points of (4.33) are not too far from the saddle points the stationary phase methods can be used as a guide.

The complex trajectory method outlined in this section has been developed by many authors. The first applications to heavy-ion scattering problems were made by Knoll & Schaeffer (1976, 1977) and by Koeling & Malfliet (1975). This method is more accurate than the simpler stationary phase method (4.34), but it is more complicated to use for numerical calculations. This is because of the additional numerical difficulties involved with solving equation (4.25) for the complex stationary points and calculating $\psi(\lambda_s)$ for complex $\lambda_s$. Because of these problems very few numerical calculations have ever been made.

In spite of these difficulties it is possible to give a physical interpretation to the angular momenta of the complex trajectories. Let $\lambda_s(\theta)$ be a complex solution of the saddle point equation (4.25). Equation (4.35) can be generalized to give

$$(\partial\psi/\partial\theta)_E = \pm\lambda(\theta)$$

The contribution of the saddle point $\lambda(\theta)$ to the amplitude $f(\theta)$ is

$$\sqrt{\sigma_{\mathrm{cL}}(\lambda(\theta))}\exp[i\,\mathrm{Re}\,\psi(\lambda(\theta)) - \mathrm{Im}\,\psi(\lambda(\theta))] \tag{4.44}$$

Equation (4.44) shows that the variations of the phase and magnitude of a saddle point contribution to the scattering amplitude with scattering angle $\theta$ are determined mainly by the real and imaginary parts of $\lambda(\theta)$ respectively. There is also some variation coming from $\sigma_{\mathrm{cL}}(\lambda(\theta))$, but in most cases this is less important. If there are two interfering amplitudes and $\mathrm{Re}\,\lambda_1$ and $\mathrm{Re}\,\lambda_2$ vary slowly with angle then the spacing between peaks in the interference structure is approximately

$$\Delta\theta \simeq 2\pi/|\mathrm{Re}(\lambda_1 - \lambda_2)| \tag{4.45}$$

or

$$\Delta\theta \simeq 2\pi/|\mathrm{Re}(\lambda_1 + \lambda_2)| \tag{4.46}$$

The relation (4.45) is appropriate if $\lambda_1$ and $\lambda_2$ are both associated with either nearside or farside amplitudes. The form (4.46) should be used if one is nearside and the other is farside. It is clear from (4.46) that nearside–farside interference is associated with a small $\Delta\theta$ and rapid oscillations in an angular distribution. The contribution of $\lambda(\theta)$ to the cross-section is

$$\sigma(\theta) \simeq |\sigma_{cL}(\lambda(\theta))| \exp(-2\,\mathrm{Im}\,\psi(\lambda(\theta))) \tag{4.47}$$

The second factor in this expression gives a cross-section varying exponentially with angle and from (4.44) the shape of the exponential decay is determined by $2\,\mathrm{Im}\,\lambda(\theta)$. In a case where there is a Coulomb rainbow two saddle points contribute to the amplitude $f_0(\theta)$. The complex saddle point positions $\lambda_a(\theta), \lambda_b(\theta)$ move as the scattering angle $\theta$ is changed. When the stationary phase method is used the saddle points are real for $\theta < \theta_R$, they coalesce when $\theta = \theta_R$ and they become complex in the shadow region $\theta > \theta_R$. In the complex trajectory method the saddle points are complex for all values of $\theta$. When $\theta$ is near the rainbow angle the saddle points $\lambda_a(\theta), \lambda_b(\theta)$ may approach one another and the saddle point approximation may break down. The Airy uniform approximation (4.41) can still be used, but the quantities $\beta, u$ and $v$ which are real in the stationary phase approximation become complex.

## 4.6 Qualitative aspects of elastic cross-sections

Figure 4.6 shows a sketch of an optical model angular distribution for ($^{16}O$, $^{58}Ni$) elastic scattering at $E_{lab} = 200$ MeV. It is an example discussed in a review article by Hussein & McVoy (1984). The ratio of the elastic cross-section $\sigma(\theta)$ to the Rutherford cross-section $\sigma_R(\theta)$ is plotted against the centre-of-mass scattering angle. We shall show how the analysis of the preceding sections can be used to give a qualitative understanding of the structure of the angular distribution.

The Poisson sum formula was used in section 4.2 to write the scattering amplitude as a sum of several components (equation 4.16). In the present example only the components with $m = 0$ give a significant contribution so that

$$f(\theta) \approx f_0^-(\theta) + f_0^+(\theta) \tag{4.48}$$

The stationary phase evaluation of the amplitudes shows that $f_0^-(\theta)$ is a nearside amplitude and that $f_0^+(\theta)$ is farside in the sense of fig. 4.1. The stationary paths contributing to $f_0^-(\theta)$ have a positive deflection angle while those contributing to $f_0^+(\theta)$ have a negative deflection angle.

The curves marked $\sigma^-(\theta)$ and $\sigma^+(\theta)$ in fig. 4.6 show the nearside and farside contributions to the cross-section. For $\theta < 30°$ the nearside

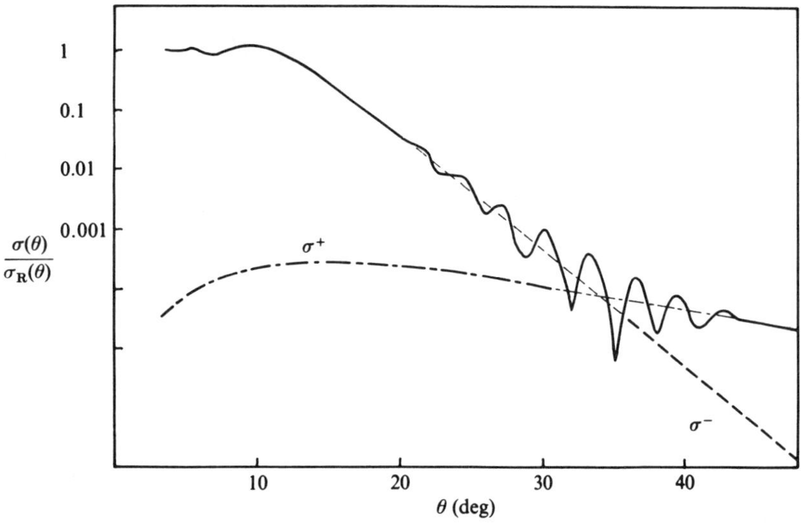


Fig. 4.6. The solid curve shows the optical model angular distribution for $^{16}O$ on $^{58}Ni$ at $E_{lab} = 200$ MeV calculated from a Woods–Saxon potential with $V = 100$ MeV, $W = 50$ MeV, $R_v = R_w = 5.87$ fm, $a_v = a_w = .64$ fm. The dashed curves marked $\sigma^-$ and $\sigma^+$ show the nearside and farside contributions.

amplitude gives the dominant contribution while the farside component dominates for $\theta > 40°$. When $\theta \approx 35°$ the nearside and farside amplitudes have comparable magnitudes and the oscillations in the cross-section near 35° are due to nearside–farside interference. The spacing between the oscillations is $\Delta\theta \approx 2.7°$ so that the real parts of the saddle point positions, which are related to $\Delta\theta$ by equation (4.46), satisfy

$$\tfrac{1}{2}|\mathrm{Re}(\lambda_1 + \lambda_2)| \approx \lambda_g = \pi/\Delta\theta \approx 65 \tag{4.49}$$

The nearside amplitude $f_0^-(\theta)$ shows a Coulomb rainbow structure near 15° and when $\theta < \theta_R$ the amplitude can be approximated by a sum of two stationary phase contributions (equation (4.39))

$$f_0^-(\theta) \simeq f_a^-(\theta) + f_b^-(\theta) \tag{4.50}$$

which interfer to give the oscillations in the cross-section at forward angles. The dominant amplitude is $f_b^-(\theta)$ which goes over into the Rutherford amplitude at forward angles, but the second component $f_a^-(\theta)$ due to deflection by the nuclear forces is important near the rainbow angle. The structure near $\theta = 10°$ is called a Fresnel diffraction pattern in Frahn's diffraction theory discussed in chapter 5. The diffraction theory also gives a decomposition like (4.50), but the interpretation of the component $f_a^-(\theta)$ is different.

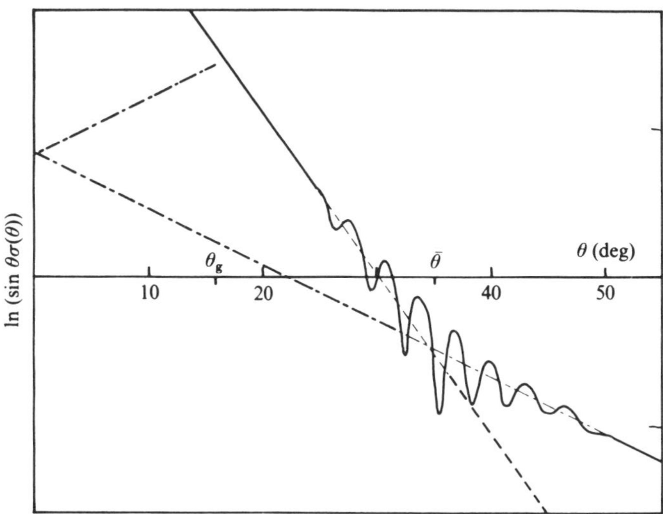


Fig. 4.7. Logarithmic plot of $\sin\theta\sigma(\theta)$ against $\theta$ for the same cross-section as fig. 4.6. The dot-dashed lines indicate the contributions of the component amplitudes $f_a^-, f_a^+$ and the dashed line of $f_d^-$ (equations (4.52) and (4.53)).

Figure 4.7 shows the same cross-section in the form of a logarithmic plot of $\sin\theta\sigma(\theta)$ against $\theta$ rather than the customary $\sigma(\theta)/\sigma_R(\theta)$ in fig. 4.6. This kind of plot was first used by Rowley & Marty (1976*a*) and subsequently by Hussein and McVoy (1984). Its advantage is that any exponential fall-off with angle of the kind predicted by equation (4.47), by Ericson's parametrization (equation 5.32) or by the Regge pole formulae of section 6.2 shows up clearly as a straight line. The diagram also shows the nearside $\sigma_a^- = |f_a^-|^2$ and farside $\sigma_a^+ = |f_a^+|^2$ deflected contributions. It is clear that $\sigma_a^+(\theta)$ is a continuation $\sigma_a^-(\theta)$ to negative angles (fig. 4.1). In the same way the amplitude $f_a^+(\theta)$ is a continuation of $f_a^-(\theta)$, but the stationary phase formula (4.29) shows that there is an additional phase factor of $\exp(-\frac{1}{2}i\pi)$. Such phase changes are characteristic features of all semi-classical theories and occur even when there is a turning point or a focal point. An example of a phase change at a turning point is the WKB connection formula (3.26). In the present case $\theta = 0$ is a focal point. The factor $(-1)^m$ in equation (4.17) has a similar origin. If an orbit (fig.4.1) turns around the nucleus once it crosses one focal point at $\theta = 0$ and another at $\theta = \pi$ and there is a phase factor of $\exp(-i\pi) = -1$. If the orbit circulates $m$ times there is a factor $(-1)^m$. In general such a phase factor has the form $\exp(-\frac{1}{2}\nu i\pi)$ where $\nu$ is an integer called the Morse or Maslov index. There is a general discussion of such indices in section 8.4.

In the stationary phase method they are contained in the terms $\chi$ in formula (4.28) for the phase $\alpha(\theta)$.

When $\theta > \theta_R$ there is a wave which spreads into the rainbow shadow region and is approximated by a saddle point contribution as discussed in section 4.4

$$f_{\bar{o}}^{-}(\theta) \approx f_{\bar{d}}^{-}(\theta) \tag{4.51}$$

Near the rainbow angle the Airy uniform approximation can be used to interpolate between the stationary phase amplitude (4.50) and the saddle point result (4.51).

The important qualitative conclusion of this chapter is that the stationary phase or saddle point approximations allow the scattering amplitude to be written as a sum of several components. In the present case as illustrated in fig. 4.7.

$$f(\theta) \approx f_{\mathrm{a}}^{+}(\theta) + f_{\mathrm{a}}^{-}(\theta) + f_{\mathrm{b}}^{-}(\theta); \quad \theta < \theta_R \tag{4.52}$$

$$f(\theta) \approx f_{\mathrm{a}}^{+}(\theta) + f_{\bar{d}}^{-}(\theta); \quad \theta > \theta_R \tag{4.53}$$

Oscillations in the cross-section are due to interference between the various components.

Chapter 5 presents a decomposition of the amplitude due to Frahn which is based on a diffraction picture. The physical ideas in the stationary phase picture are different from the ones in the diffraction picture, but the general structure of the decomposition given in equations (4.52) and (4.53) is the same. All the semi-classical approaches agree that the scattering amplitude can be decomposed into various components. There is some disagreement about the correct physical description of the components.

## 4.7 Fuller's nearside–farside decomposition

The nearside–farside decomposition of the elastic scattering amplitude was studied in the early part of this chapter by using the Poisson sum formula to write $f(\theta)$ as a sum of integrals and then evaluating the integrals by the stationary phase approximation. Fuller (1975) has shown that it is possible to make the decomposition in another way by working directly with the partial wave series (2.17). This decomposition is achieved by replacing the Legendre polynomial by its travelling wave components $Q_l^{+}(x)$ and $Q_l^{-}(x)$ respectively

$$P_l(x) = Q_l^{+}(x) + Q_l^{-}(x) \tag{4.54}$$

where for large $l$ and $l^{-1} < \theta < \pi - l^{-1}$

$$Q_l^{\pm}(\cos\theta) \sim (2\pi\lambda \sin\theta)^{-\frac{1}{2}} \exp[\pm i\lambda\theta - \tfrac{1}{4}i\pi] \tag{4.55}$$

Comparison with equation (4.17) shows that this is the required behaviour for a nearside–farside decomposition. The travelling wave functions are defined in terms of the $P_l$ and the Legendre functions of the second kind $Q_l$ by

$$Q_l^{\pm}(x) = \tfrac{1}{2}(P_l(x) \mp (2i/\pi)Q_l(x)) \tag{4.56}$$

Fuller defines the nearside component $f^-(\theta)$ and the farside component $f^+(\theta)$ by

$$f^{\pm}(\theta) = f_C^{\pm}(\theta) + 1/(2ik)\sum_l (2l+1)\exp(2i\sigma_l)(S_{nl}-1)Q_l^{\mp}(\cos\theta) \tag{4.57}$$

where $f_C^{\pm}(\theta)$ are the corresponding components of the Coulomb amplitude. The total amplitude is

$$f(\theta) = f^-(\theta) + f^+(\theta) \tag{4.58}$$

and (4.57) is exactly equivalent to the partial wave series (2.17).

The most difficult part of the amplitude (4.57) is the Coulomb term. Simple classical considerations would suggest that repulsive Coulomb scattering has only a nearside contribution and that the farside part is zero. This is almost the case. For forward angles $f_C^-(\theta) \gg f_C^+(\theta)$. On the other hand, the Coulomb amplitude is regular for $\theta \simeq \pi$, the nearside and farside components have comparable magnitudes at backward angle and it is necessary to use the correct components $f_C^{\pm}(\theta)$. These have been obtained by Fuller (1975) in closed form and are given in his paper.

The decomposition (4.57) has two very nice features:

(a) it is exact,
(b) it is relatively easy to include as a part of a standard optical model code.

Hussein & McVoy (1984) have studied a number of examples of nearside–farside decompositions in a review article on 'practical heavy-ion optics'. They write the elastic cross-section as

$$\sigma(\theta) = \sigma^-(\theta) + \sigma^+(\theta) + 2(\sigma^-\sigma^+)^{\frac{1}{2}}\cos\phi(\theta) \tag{4.59}$$

where $\sigma^{\pm}(\theta) = |f^{\pm}(\theta)|^2$ and $\phi(\theta)$ is the phase difference between the nearside and farside amplitudes. They show that the rapid oscillations in $\sigma(\theta)$ with a spacing $\Delta\theta \simeq \pi/l_g$ are a consequence of the nearside–farside interference as discussed in section 4.6. The cross-section components $\sigma^+(\theta)$ and $\sigma^-(\theta)$ vary smoothly with angle as expected from the semi-classical analysis. Hussein & McVoy point out that the component amplitudes $f^+(\theta), f^-(\theta)$ are defined by the partial wave series (4.57) and can only be obtained from the partial wave amplitudes. No rigorous way is known for decomposing the angular distribution in nearside–farside form without going through the partial wave analysis.

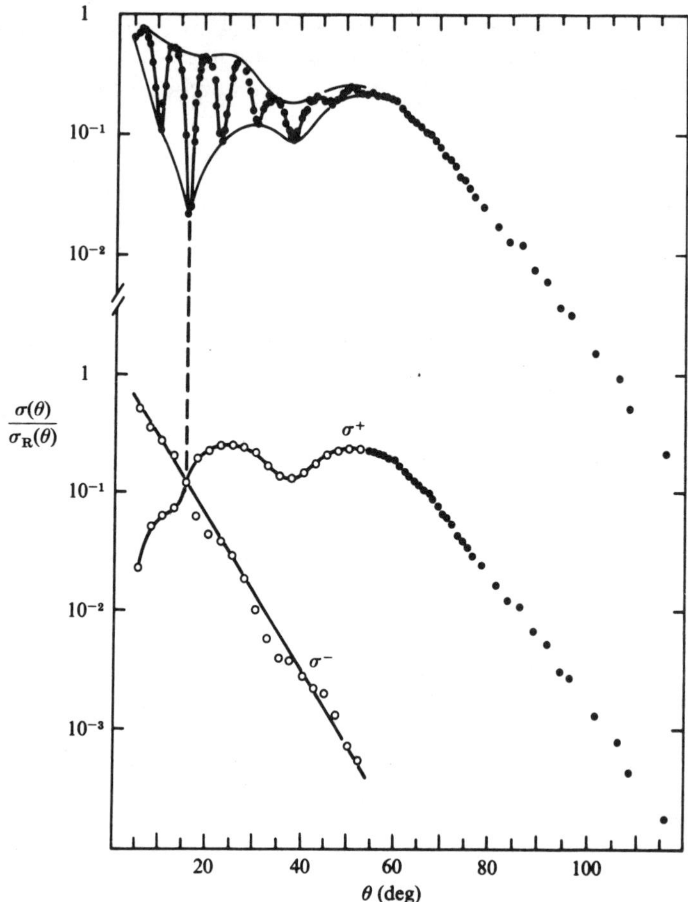


Fig. 4.8. Elastic scattering of α-particles on $^{40}$Ca at $E_{\text{lab}} = 104$ MeV. The upper part of the diagram shows the experimental data and the envelopes $E_+$ and $E_-$. The lower part of the diagram shows the nearside and farside components $\sigma^-(\theta)$ and $\sigma^+(\theta)$. The structure in $\sigma^+(\theta)$ is an example of a nuclear rainbow effect.

Although this is true from a strictly mathematical point of view da Silveira & Leclercq–Willain (1984) have shown that, in some cases, an approximate phenomenological decomposition can be made directly from the experimental cross-section. Their analysis starts from the observation that the nearside–farside interference produces regular rapid oscillations in the cross-section with spacing $\Delta\theta\pi/l_g$. The envelopes of the oscillatory pattern from equation (4.59) are

$$\left.\begin{aligned} E_+(\theta) &= (\sqrt{\sigma^+(\theta)} + \sqrt{\sigma^-(\theta)})^2 \\ E_-(\theta) &= (\sqrt{\sigma^+(\theta)} - \sqrt{\sigma^-(\theta)})^2 \end{aligned}\right\} \tag{4.60}$$

Such envelopes are easy to draw whenever the extrema of the oscillatory pattern are well defined in the experimental data. If so, it becomes straight forward to obtain $\sigma^-(\theta)$ and $\sigma^+(\theta)$ at each scattering angle. A useful relation for this purpose is

$$\sigma^{\pm}(\theta)/E_+(\theta) = \tfrac{1}{4}(1 \pm (E_-(\theta)/E_+(\theta))^{\frac{1}{2}})^2 \qquad (4.61)$$

The positive sign in (4.61) gives the larger of $\sigma^-(\theta)$ and $\sigma^+(\theta)$ at each angle and the negative sign gives the smaller of the two. Some additional argument is needed to decide which is which. Da Silveira *et al.* apply this procedure to two examples. The results of one of them, the elastic scattering of $\alpha$ on $^{40}$Ca at $E_{\text{lab}} = 104$ MeV, is shown in fig. 4.8. The upper part of the diagram shows the experimental data and the envelopes $E_+$ and $E_-$. At each angle (4.60) is solved to give $\sigma^+(\theta)$ and $\sigma^-(\theta)$. The results are shown by open circles in the lower part of the diagram. The additional argument used to draw the curves is that the nearside cross-section $\sigma^-(\theta)$ is associated with the amplitude $f_d^-(\theta)$ (equation 4.51) on the shadow side of the Coulomb rainbow. Thus $\sigma^-(\theta)$ approaches the Rutherford cross-section $\sigma_R(\theta)$ for small $\theta$ and decreases exponentially as $\theta$ increases. The discussion of fig. 4.18 is continued in the next section.

## 4.8 The nuclear rainbow

The cross-section shown in fig. 4.8 is an example of nuclear rainbow scattering. This is a refractive effect which manifests itself in the farside part of the cross-section $\sigma^+(\theta)$. There is a pronounced bump in $\sigma^+(\theta)$ at $\theta \simeq 55°$ beyond which the cross-section decreases almost exponentially. In the semi-classical picture developed in section 4.3 the nuclear rainbow is associated with a classical deflection function like the one shown in fig. 4.9. The attractive nuclear force produces a negative deflection for small angular momenta. There are two rainbow angles where $d\Theta/d\lambda = 0$. The outer one with rainbow angle $\Theta_{CR}$ is the Coulomb rainbow and corresponds to crossing the Coulomb barrier. The inner rainbow at $\Theta_{nR}$ is called a nuclear rainbow because it is associated with a trajectory which depends on the nuclear force at smaller distances. Because a nuclear rainbow corresponds to a negative deflection it appears as a modification of the farside component of the scattering amplitude.

The following information can be extracted from fig. 4.8.

(i) The farside cross-section $\sigma^+(\theta)$ shows a characteristic Airy interference pattern which was discussed in section 4.4 and illustrated in fig. 4.5. If the scattering angle is forward of the rainbow angle two stationary angular momenta ($\lambda_1$ and $\lambda_2$ in fig. 4.9) contribute to the cross-section.

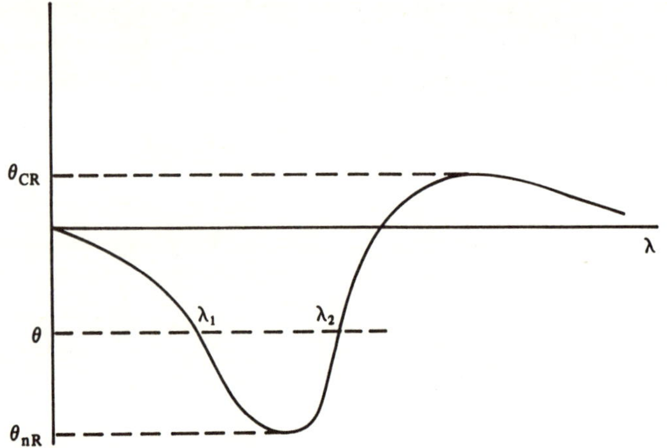


Fig. 4.9. Nuclear rainbow deflection function. $\theta_{CR}$ and $\theta_{nR}$ are the Coulomb and nuclear rainbow angles. $\lambda_1$ and $\lambda_2$ are the stationary angular momenta contributing to the cross-section at the angle $\theta$.

The corresponding amplitudes interfere to produce the oscillations in $\sigma^+(\theta)$ which are seen in fig. 4.8. Equation (4.45) shows that the measured spacing between the peaks of the pattern ($\Delta\theta \simeq 25°$) gives an indication of the difference between $\lambda_2$ and $\lambda_1$

$$\lambda_2 - \lambda_1 \approx 2\pi/\Delta\theta \approx 15$$

Beyond the rainbow angle at $\Theta_{nR} \approx 60°$ the cross-section drops almost exponentially. This is the rainbow shadow, the stationary angular momentum $\lambda(\theta)$ is complex and the slope (equation (4.47)) gives a measure of Im $\lambda(\theta)$.

(ii) The cross-section $\sigma^-(\theta)$ shows the exponential decrease characteristic of a Coulomb rainbow in the angular region beyond the Coulomb rainbow angle. Again the slope of the exponential measures the imaginary part of the saddle point position.

(iii) The cross-sections $\sigma^-(\theta)$ and $\sigma^+(\theta)$ cross for $\theta \approx 15°$. This is near the angle where the interference pattern between the near and farside amplitudes shows a very deep minimum. Such a minimum appears in all $\alpha$-nucleus elastic angular distributions at intermediate energies. The spacing $\Delta\theta \approx 7°$ of the nearside–farside interference oscillations in the upper part of fig. 4.18 can be used to estimate $\lambda_{CR} + \lambda_{nR}$ from equation (4.46)

$$\lambda_{CR} + \lambda_{nR} \approx 2\pi/\Delta\theta \approx 50$$

The appearance of a nuclear rainbow is characteristic of a relatively weak absorption. It was observed in the scattering of $\alpha$-particles by many nuclei at energies $E_{lab} \gtrsim 80\,MeV$ (Goldberg & Smith, 1974; Goldberg *et al.*, 1974; Goldberg, 1975). There is also clear evidence in $^6$Li scattering from $^{12}$C and $^{40}$Ca (Cook *et al.*, 1982) and there are also some indications of a nuclear rainbow in $^9$Be elastic scattering from $^{12}$C and $^{16}$O (Satchler *et al.*, 1983) and possibly for ($^{12}$C, $^{12}$C) scattering at 300 MeV (Branden, 1982).

# 5

# Frahn's diffraction theory

## 5.1 The strong absorption model

The strong absorption model (SAM) was developed by Frahn & Venter (1963) and was a refinement of the 'sharp cut-off model' proposed by Blair (1954). Blair's model makes the most extreme assumption (2.23) about strong absorption in $l$-space, namely that all partial waves with angular momentum less than a grazing angular momentum $l_g$ are completely absorbed, while higher partial waves are affected by the Coulomb interaction only. It was found empirically and by numerical calculations that the scattering cross-section at an angle $\theta_g$ related to $l_g$ by Rutherford's formula (2.32) is approximately one-quarter of the Rutherford cross-section

$$\sigma(\theta_g) \approx \tfrac{1}{4}\sigma_R(\theta_g) \tag{5.1}$$

Blair's model reproduced some of the main features in the angular distributions for elastic scattering of heavy-ions in cases where the Coulomb repulsion is strong in the sense that the Sommerfeld parameter $n$ is large. In Blair's model oscillations in the angular distributions clearly have a diffractive origin due to the sharp cut-off in the nuclear partial wave amplitudes.

Frahn & Venter generalized Blair's model by allowing for a smooth cut-off in the partial wave amplitude and by including the effects of a real nuclear phase. They also introduced a number of analytic devices (Venter, 1963) including the Poisson sum formula and the replacement of Legendre polynomials by their asymptotic forms which enabled them to obtain closed form expressions for partial wave amplitudes. Frahn and his collaborators developed the strong absorption model over a number of years by using improved parametrization for the partial wave amplitudes and by inventing more sophisticated analytical methods.

The strong absorption model of Frahn can be used when the magnitude of the nuclear partial wave amplitude has the form shown in fig. 5.1. We use an amplitude

$$S_n(\lambda) = \exp(2i\delta_n(\lambda)) \tag{5.2}$$

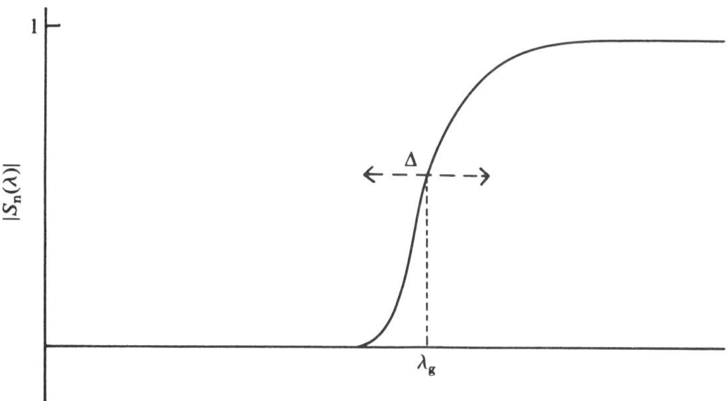


Fig. 5.1. Typical magnitude of the $S$-matrix element $|S_n(\lambda)|$ plotted against $\lambda = l + \frac{1}{2}$ for the strong absorption model (SAM). The grazing angular momentum is denoted by $\lambda_g$.

which is a function of the continuous angular momentum $\lambda = l + \frac{1}{2}$ as discussed in chapters 3 and 4. The width $\Delta$ of the transition region around the grazing angular momentum $\lambda_g$ should be sufficiently small. The basic idea is that there are two main contributions to the amplitude $f(\theta)$,

(a) the semi-classical Coulomb contribution,
(b) a diffractive contribution from the region $\Delta$ around the grazing angular momentum $\lambda_g = n \cot\frac{1}{2}\theta_g$.

In the following sections we put

$$|\Theta'_g| = |\partial\Theta_C/\partial\lambda_g| = 2n/(n^2 + \lambda_g^2) = (2/n)\sin^2\tfrac{1}{2}\theta_g \tag{5.3}$$

where $\Theta$ is the Coulomb scattering angle. For Frahn's approach to be useful it is necessary that the width $\Delta$ of the cut-off in $S_n(\lambda)$ should be small in the sense

$$\tfrac{1}{2}|\Theta'_g|\Delta^2 \ll 1 \tag{5.4}$$

This condition was given by Frahn & Gross (1976). A physical argument behind it is that there are two angular spreads associated with $\Delta$. One is the range $(\Delta\theta_C = |\Theta'_g|\Delta)$ of the classical Coulomb scattering angles associated with the width $\Delta$ in fig. 5.1. The other is a diffraction spread $(\Delta\theta)_d \approx 1/\Delta$. Frahn's method can be used when the diffraction spread dominates $(\Delta\theta)_d \gg (\Delta\theta)_C$ which is equivalent to the condition (5.4).

## 5.2 The diffraction theory

The sharp cut-off amplitude (2.23) leads to a Poisson series for $f(\theta)$ which converges very slowly. The term with $m = 0$ gives the dominant contribution for small $\theta$, but many terms have to be retained to get an accurate result. In contrast the strong absorption model yields a rapidly converging Poisson series because the transition region in the partial wave amplitude $S_n(\lambda)$ has non-zero width $\Delta$. Normally only the terms $f_0^{\pm}(\theta)$ need be used. In this section we give Frahn's method for calculating these amplitudes. We present only the simplest version of his theory. A more complete treatment can be found in Frahn (1984).

The amplitudes $f_0^{\pm}(\theta)$ are given by the integrals (4.17)

$$f_0^{\pm}(\theta)=\frac{1}{ik}(2\pi\sin\theta)^{-\frac{1}{2}}\int_0^{\infty}\lambda^{\frac{1}{2}}d\lambda S_n(\lambda)\exp i(2\sigma(\lambda)\pm\lambda\theta\mp\tfrac{1}{4}\pi) \qquad (5.5)$$

where $S_n(\lambda)=\exp(2i\delta_n(\lambda))$ is the semi-classical nuclear partial wave amplitude and $\sigma(\lambda)$ is the Coulomb phase (3.47). The assumption of the strong absorption model is that $S_n(\lambda)$ has a relatively sharp cut-off near $\lambda_g$ as illustrated in fig. 5.1. The width of the cut-off is $\Delta$ and

$$\begin{aligned} S_n(\lambda) &\simeq 1 \quad \text{if} \quad \lambda>\lambda_g+\Delta \\ &\simeq 0 \quad \text{if} \quad \lambda<\lambda_g-\Delta \end{aligned} \qquad (5.6)$$

The basic idea is that the main contributions to the integral (5.5) come either from stationary points of the Coulomb amplitude or from the region near $\lambda_g$.

First we state the results and then discuss their derivation. Approximate formulae for the nearside ($f_0^-$) and farside ($f_0^+$) amplitudes are

(a) $\theta>\theta_g$; nearside

$$f_0^-(\theta)\simeq f_C(\theta_g)\left(\frac{\sin\theta_g}{\sin\theta}\right)^{\frac{1}{2}}J(\theta_g-\theta)\exp\{i\lambda_g(\theta_g-\theta)\} \qquad (5.7)$$

(b) $\theta<\theta_g$; nearside

$$f_0^-(\theta)\simeq f_C(\theta)-f_C(\theta_g)\left(\frac{\sin\theta_g}{\sin\theta}\right)^{\frac{1}{2}}J(\theta_g-\theta)\exp\{i\lambda_g(\theta_g-\theta)\} \qquad (5.8)$$

(c) farside

$$f_0^+(\theta)\simeq if_C(\theta_g)\left(\frac{\sin\theta_g}{\sin\theta}\right)^{\frac{1}{2}}J(\theta_g+\theta)\exp\{i\lambda_g(\theta_g+\theta)\} \qquad (5.9)$$

In these equations $f_C(\Theta)$ is the semi-classical Rutherford amplitude and the functions $J$ are

(a) $\chi<0$

$$J(\chi)=\left(\frac{|\Theta_g'|}{2\pi}\right)^{\frac{1}{2}}e^{\frac{1}{4}i\pi}\int_{-\lambda_g}^{\infty}d\mu S_n(\lambda_g+\mu)\exp i(\chi\mu-\tfrac{1}{2}|\Theta_g'|\mu^2) \qquad (5.10)$$

(b) $\chi > 0$

$$J(\chi) = \left(\frac{|\Theta'_g|}{2\pi}\right)^{\frac{1}{2}} e^{\frac{1}{4}i\pi} \int_{-\lambda_g}^{\infty} d\mu (1 - S_n(\lambda_g + \mu)) \exp i(\chi\mu - \tfrac{1}{2}|\Theta'_g|\mu^2) \tag{5.11}$$

The derivation of the formula for $f_0^-(\theta)$ in case (b) when $\theta < \theta_g$ runs as follows: the integral in equation (5.5) is written as a sum of two parts $I = I_C + I_1$ where

$$I_C(\theta) = \int_0^{\infty} \lambda^{\frac{1}{2}} d\lambda \exp i(2\sigma(\lambda) - \theta\lambda + \tfrac{1}{4}\pi)$$

$$I_1(\theta) = \int_0^{\infty} \lambda^{\frac{1}{2}} d\lambda \exp i(2\sigma(\lambda) - \theta\lambda + \tfrac{1}{4}\pi)(S_n(\lambda) - 1)$$

The first integral $I_C$ occurs in the semi-classical formula for Coulomb scattering and was evaluated in section 4.3. The second integral $I_1$ has no stationary point in the region $0 < \lambda \lesssim \lambda_g + \Delta$ where $S_n - 1$ is non-zero and according to Frahn's argument the main contribution comes from values of $\lambda$ near $\lambda_g$. The integral is approximated by

(a) replacing $\lambda^{\frac{1}{2}}$ by $\lambda_g^{\frac{1}{2}}$,
(b) expanding the Coulomb phase as a power series in $(\lambda - \lambda_g)$ up to second order

$$2\sigma(\lambda) \approx 2\sigma(\lambda_g) + \theta_g(\lambda - \lambda_g) + \tfrac{1}{2}\Theta'_g(\lambda - \lambda_g)^2 \tag{5.12}$$

Terms in the resulting expression can be collected to give

$$I_1(\theta) = -I_C(\theta_g) J(\theta_g - \theta) \tag{5.13}$$

with $J(\chi)$ given by (5.11). If these results are substituted into equation (5.5) the term $I_C(\theta)$ gives $f_C(\theta)$ and the term $I_1$ gives the second term in (5.8). The other two cases (a) and (c) can be evaluated in a similar way. In these cases the integral (5.5) has no stationary points for $\lambda > \lambda_g + \Delta$ and there are contributions only from $\lambda$-values near the cut-off at $\lambda_g$ in $S_n(\lambda)$.

When the condition (5.4) holds the function $J(\chi)$ can be expressed in terms of the Fourier transform of the derivative of the nuclear scattering amplitude

$$F(\chi) = \int_0^{\infty} d\lambda (dS_n/d\lambda) \exp\{i\chi(\lambda - \lambda_g)\} \tag{5.14}$$

as

$$J(\chi) \simeq 2^{-\frac{1}{2}} e^{-\frac{1}{4}i\pi} F(\chi)(f(|w|) + ig(|w|)) \tag{5.15}$$

$$w = \chi(\lambda_g/\pi \sin\theta_g)^{\frac{1}{2}} = \chi/(\pi|\theta'_g|)^{\frac{1}{2}} \tag{5.16}$$

The functions $f(w)$ and $g(w)$ are auxiliary functions for Fresnel integrals

(F.1). They have a simple rational approximation (F.2) which makes them easy to calculate. When $\chi = 0$, $F(0) = 1$ and $J(0) = \frac{1}{2}$. When $w$ is large

$$f(w) \approx 1/(\pi w), \quad g(w) \simeq 0 \tag{5.17}$$

The result (5.15) will be derived in appendix F. The derivation depends on the fact that $(dS_n/d\lambda)$ is a sharply peaked function of $\lambda$ centered at $\lambda_g$ with width $\Delta$. It follows that $F(\chi)$ is a slowly varying function of $\chi$ with width $1/\Delta$.

Equations (5.7), (5.8) and (5.17) ensure that $f_0(\theta)$ is continuous at $\theta = \theta_g$. At $\theta_g$

$$f_0^-(\theta_g) = \tfrac{1}{2} f_C(\theta_g)$$

Thus, when the component $f_0^+$ is negligible at the grazing angle the theory gives the one-quarter point recipe (5.1) of Blair.

### 5.3 The parametrized *S*-matrix

The central theoretical object in the work of Frahn and his collaborators is the set of partial wave amplitudes $S_l$ and its extension $S(\lambda)$ to continuous angular momenta. Frahn (1984) calls the $S_l$ the elements of the partial wave *S*-matrix. It is regarded as being an intermediary between the experimental cross-section on the one hand and the underlying many-body hamiltonian or to an effective interaction such as the optical model potential on the other. The fundamental property of $S(\lambda)$ is the strong absorption character and in much of their work Frahn and his co-workers have used reasonable parametrized forms for $S(\lambda)$ rather than tried to relate it to an underlying optical potential.

We give three simple parametrized forms for the nuclear partial wave amplitude:

(a) $$S_n(\lambda) = g\left(\frac{\lambda - \lambda_g}{\Delta}\right) + i\alpha\Delta \frac{d}{d\lambda} g\left(\frac{\lambda - \lambda_g}{\Delta}\right) \tag{5.18}$$

$$g(x) = (1 + e^{-x})^{-1}$$

(b) $$S_n(\lambda) = [1 + \exp((\lambda_g - \lambda)/\Delta - i\alpha)]^{-1} \quad 0 < \alpha < \tfrac{1}{2}\pi \tag{5.19}$$

(c) $$S_n(\lambda) = \exp(2i\delta_n(\lambda))$$

$$2\delta_n(\lambda) = \exp[(\lambda_g - \lambda)/\tilde{\Delta} + i(\tfrac{1}{2}\pi - \tilde{\alpha})] \quad 0 < \tilde{\alpha} < \tfrac{1}{2}\pi \tag{5.20}$$

An amplitude similar to (a) was used in the original work of Frahn & Venter (1964). The form (b) was introduced by Ericson (1966) and has been used by Rowley & Marty (1976*a*), by Frahn (1984) and others. Version (c) is a simplified form of an amplitude suggested by Kaufmann (1977). Kaufmann's expression (5.20) is interesting since it coincides with the semi-classical perturbation expression (3.68) for the nuclear phase shift calculated with an exponential potential. It therefore provides a link between the semi-classical phase and the parametrized amplitude approaches. There is also a connection between the parameters $\tilde{\alpha}$ and $\tilde{\Delta}$ in (5.20) and the optical model parameters of the exponential potential

$$\tilde{\alpha} = \tan^{-1}(V_0/W_0), \quad \tilde{\Delta} = ka(n^2 + \lambda_g^2)^{\frac{1}{2}}/\lambda_g \tag{5.21}$$

where $V_0$ and $W_0$ are the real and imaginary strength parameters and $a$ is the surface diffuseness.

The three amplitudes are written in a way which facilitates comparisons between them. In all cases $\alpha = 0$ corresponds to a purely absorptive amplitude ($S_n$ is real). The Frahn–Venter amplitude can be obtained from Ericson's form by expanding it to first order in $\alpha$. The expressions (b) and (c) coincide for $\lambda - \lambda_g \gg \Delta$ if $\alpha = \tilde{\alpha}$ and $\Delta = \tilde{\Delta}$.

The Fourier transforms (5.14) of the derivatives of the three amplitudes can be calculated in closed form and are

$$\text{(a)} \quad F(\chi) = \frac{\pi\Delta\chi}{\sinh(\pi\Delta\chi)}(1 + \alpha\Delta\chi) \tag{5.22a}$$

$$\text{(b)} \quad F(\chi) = \frac{\pi\Delta\chi}{\sinh(\pi\Delta\chi)}\exp(\alpha\Delta\chi) \tag{5.22b}$$

$$\text{(c)} \quad F(\chi) = \exp(\tilde{\alpha}\tilde{\Delta}\chi)\Gamma(1 - i\tilde{\Delta}\chi)$$

$$= \exp(i\gamma(\chi))\exp(\tilde{\alpha}\tilde{\Delta}\chi)\left(\frac{\pi\tilde{\Delta}\chi}{\sinh(\pi\tilde{\Delta}\chi)}\right)^{\frac{1}{2}} \tag{5.22c}$$

where $\gamma(\chi)$ is a phase from the gamma function.

Ericson's parametrization (5.19) of $S_n(\lambda)$ has a line of poles at

$$\lambda = \lambda_g + ((2n + 1)\pi - \alpha)\Delta i$$

where $n$ is an integer. Those with $n > 0$ lie in the upper half of the complex $\lambda$-plane and could correspond to Regge (section 6.2) or Sommerfeld poles (section 6.4). The poles with $n < 0$ lie in the lower half of the $\lambda$-plane. A nuclear amplitude calculated from a potential cannot have poles in the lower part of the $\lambda$-plane so that those poles in Ericson's $S_n(\lambda)$ are an artefact of the parametrization. They give a simple representation of a more complex structure in the true $S_n(\lambda)$. The Fourier transform $F(\chi)$ can be calculated by a contour integral. When $\pi\Delta\chi \gtrsim 1$ the contour can be closed in the upper half-plane and the dominant contribution comes from the pole at $\lambda = \lambda_g + (\pi - \alpha)\Delta i$ which is nearest to the real $\lambda$-axis. The pole contribution gives

$$F(\chi) \approx 2\pi\Delta\chi \exp\{-(\pi - \alpha)\Delta\chi\} \tag{5.23}$$

When $\pi\Delta\chi \lesssim -1$ the contour can be closed in the lower half-plane and the nearest pole at $\lambda = \lambda_g - (\pi + \alpha)\Delta i$ gives

$$F(\chi) \simeq -2\pi\Delta\chi \exp\{(\pi + \alpha)\Delta\chi\} \tag{5.24}$$

These formulae are consistent with (5.22b). The contour method will be used to evaluate other integrals in later chapters. Ericson's parametrization should give a good description of the barrier wave amplitude discussed in section 6.3. The pole at $\lambda_g + (\pi - \alpha)\Delta i$ corresponds to the nearest Regge pole (Sommerfeld pole) (6.29) associated with orbiting and $\alpha$ can be determined by fitting the forward angle oscillations (see equation (5.34)). This point will be discussed further in section 5.6.

## 5.4 Formulae for cross-sections

The explicit formulae (5.7)–(5.9) for the amplitudes $f_0^{\pm}(\theta)$ yield simple closed form expressions for cross-sections in some special cases. When the nearside amplitude dominates and the farside component is negligible the ratio of the elastic to the Rutherford cross-section resembles a Fresnel diffraction pattern. For $\theta < \theta_g$

$$\sigma(\theta)/\sigma_R(\theta) \approx (1 - 2A(\theta)\cos\xi(\theta) + A^2(\theta)) \tag{5.25}$$

$$A(\theta) = \left(\frac{\sigma_R(\theta_g)\sin\theta_g}{\sigma_R(\theta)\sin\theta}\right)^{\frac{1}{2}} |J(\theta_g - \theta)|$$

$$\xi(\theta) = 2n\ln\left(\frac{\sin\frac{1}{2}\theta}{\sin\frac{1}{2}\theta_g}\right) + \lambda_g(\theta_g - \theta) + \arg J$$

For $\theta > \theta_g$

$$\sigma(\theta)/\sigma_R(\theta) \approx \left(\frac{\sigma_R(\theta_g)\sin\theta_g}{\sigma_R(\theta)\sin\theta}\right)|J(\theta_g - \theta)|^2 \tag{5.26}$$

Equation (5.25) is obtained from (5.8). The function $J(\theta_g - \theta)$ is given in (5.15) and is proportional to the Fourier transform $F$ of $(dS_n/d\lambda)$. The first term in the phase $\xi(\theta)$ comes from the phase of the Coulomb amplitude (equation (2.5)).

Frahn's Fresnel formula (2.24) is a special case when there is a sharp cut-off so that $F = 1$ and when $(\theta_g - \theta)$ is small enough for the phase $\xi(\theta)$ to be approximated by expanding it as a power series in $(\theta_g - \theta)$ up to quadratic terms

$$\xi(\theta) \approx -n(\theta_g - \theta)^2/(4\sin^2\tfrac{1}{2}\theta_g) + \arg J \tag{5.27}$$

The approximation (5.27) is not very accurate and is responsible for much of the poor quantitative agreement between the simple Fresnel formula

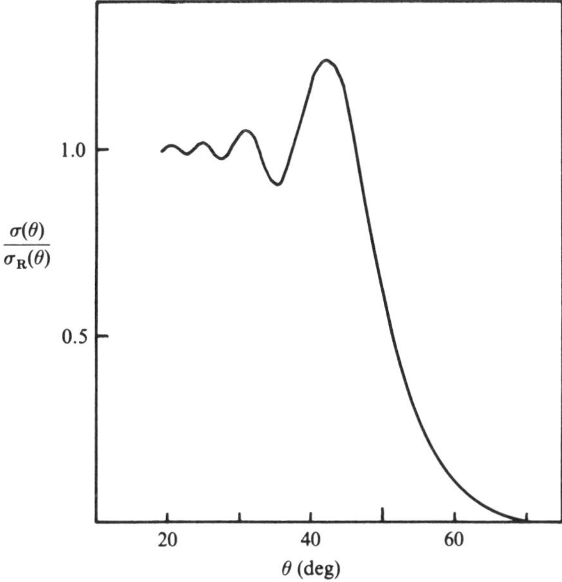


Fig. 5.2. Angular distribution for elastic scattering of $^{12}C$ by $^{208}Pb$ at $E_{lab} =$ 96 MeV calculated with Ericson's parametrization (5.19). The parameter values are $\lambda_g = 51.5$, $\alpha = .85$, $\Delta = 2.4$. The Sommerfeld parameter is $n = 27.4$ and $\theta_g = 56°$.

(2.24) and the experimental data. The two formulae (5.25) and (5.26) join continuously at $\theta = \theta_g$ and both yield Blair's quarter-point formula (5.1). Ericson's parametrization gives a function $F$ which is real (equation 5.23) and the phase of $F$ in Kaufmann's parametrization is small. These examples suggest that the phase $\xi(\theta)$ in (5.25) is determined mainly by the Coulomb terms and is not very sensitive to the nuclear part. Frahn & Venter (1963) gave a simple formula for the position $\theta_0$ of the main Fresnel peak which is based on this fact

$$\theta_0 \approx \theta_g - (3\pi/n)^{\frac{1}{2}} \sin \tfrac{1}{2}\theta_g$$

It can be obtained from (5.27) by putting $\arg J = -\frac{1}{4}\pi$, its asymptotic value when $w$ is large, and $\xi(\theta) = -\pi$, which is the condition for the main Fresnel peak.

Fig. 5.2 shows an example of an angular distribution for $^{12}C + {}^{208}Pb$ elastic scattering at $E_{lab} = 96$ MeV calculated from equations (5.25) and (5.26) with Ericson's parametrization (5.19). The parameter values are given in the figure caption. The parameter $\Delta$ measures the width of the transition region in $S_n(\lambda)$ between zero absorption for $\lambda \gg \lambda_g$ and complete absorption for $\lambda \ll \lambda_g$. The phase $\alpha$ which must lie in the range $0 < \alpha < \frac{1}{2}\pi$ simulates the effects of a real attractive nuclear potential. It gives a positive nuclear phase shift whose maximum value is $\frac{1}{2}\alpha$. The constant $|\Theta'_g|$ is a derivative of the Coulomb scattering angle (equation (5.3)) and $|\Theta'_g|\Delta^2 = 0.09 \ll 1$ so that the criterion (5.4) for the validity of Frahn's approximation is satisfied.

Rowley & Marty (1976*a*, *b*) obtained a number of very simple results using Ericson's parametrization.
When

$$|\theta_g - \theta| > 1/(\pi\Delta) \quad \text{and} \quad |\theta_g - \theta| > (\pi|\Theta'_g|)^{\frac{1}{2}}$$

the function $F$ is proportional to an exponential and the auxiliary Fresnel functions $f$ and $g$ in (5.15) can be approximated by their asymptotic forms (5.17):
When $\theta_g > \theta$

$$|J(\theta_g - \theta)| \approx (2\pi\Delta^2|\Theta'_g|)^{\frac{1}{2}} \exp[\Delta(\alpha - \pi)(\theta_g - \theta)] \tag{5.28}$$

and when $\theta_g < \theta$

$$|J(\theta_g - \theta)| \approx (2\pi\Delta^2|\Theta'_g|)^{\frac{1}{2}} \exp[\Delta(\pi + \alpha)(\theta - \theta_g)] \tag{5.29}$$

Thus in the shadow region $\theta_g < \theta$ the cross-section contains a factor which decays exponentially beyond the grazing angle with a decay constant $2(\pi + \alpha)\Delta$. In the bright region $\theta_g > \theta$ the factor $A$ in (5.25) which is proportional to $J$ determines the amplitude of the interference pattern between the Rutherford amplitude and the diffracted amplitude. The term $A(\theta)$ contains an exponential factor (5.28) with a decay constant $(\pi - \alpha)$. These formulae show that increasing $\alpha$ for fixed $\Delta$ and $\lambda_g$ reduces the cross-section in the shadow region $\theta_g < \theta$ and enhances the amplitude of the Fresnel diffraction oscillations in the bright region $\theta_g > \theta$.

These results enable us to understand the qualitative effect of the real part of the optical potential on the angular distribution. The constant $\alpha$ in Ericson's parametrization is a measure of the strength of the real part of the optical potential compared with the imaginary part ($\alpha = 0$ for a purely imaginary potential). A large $\alpha$ enhances the diffracted amplitude for $\theta_g > \theta$ and suppresses it for $\theta_g < \theta$. It appears that the attractive real potential pulls the diffracted amplitude towards forward angles.

For $\theta_g < \theta$ the scattering amplitude has contributions from both the nearside and the farside components. If these have comparable magnitudes they can interfere. The cross-section calculated from the amplitudes (5.7) and (5.9) is

$$\begin{aligned}\sigma(\theta) \approx \sigma_R(\theta_g)\left(\frac{\sin\theta_g}{\sin\theta}\right)&|J(\theta_g - \theta)|^2 \\ &\times (1 - 2b(\theta)\sin(2\lambda_g\theta + \eta(\theta)) + b^2(\theta))\end{aligned} \tag{5.30}$$

$$b(\theta) = |J(\theta_g + \theta)/J(\theta_g - \theta)|$$

$$\eta(\theta) = \arg J(\theta_g + \theta) - \arg J(\theta_g - \theta)$$

where the function $J(\chi)$ is defined in (5.15). The Fraunhofer diffraction formula (2.28) is a special case of (5.30) in the sharp cut-off model ($F = 1$) when the Coulomb field is weak ($\theta_g$ very small) and the grazing angular momentum is large. It can be obtained by assuming that $\theta_g \ll \theta$ and that $\theta$ is reasonably small so that $\sin\theta \simeq \theta$. In this limit $b(\theta) \approx 1$, $\eta(\theta) \approx 0$,

$$\sigma_R(\theta_g)(\sin\theta_g/\sin\theta) \approx \lambda_g^2/(k^2\theta_g\theta)$$

$$|J(\theta_g - \theta)|^2 \approx \theta_g/(2\pi\lambda_g\theta^2)$$

and formula (5.30) reduces to

$$\sigma(\theta) \approx [2\lambda_g/(\pi k^2\theta^3)]\sin^2(\lambda_g\theta - \tfrac{1}{4}\pi)$$

This is equivalent to the Fraunhofer diffraction cross-section (2.30) when $\lambda_g\theta \gtrsim 1$. All the formulae for amplitudes and cross-sections in this chapter are restricted to the angular region $\theta \gtrsim 1/\lambda_g$ because they are based on equation (4.17) and use the asymptotic form (4.14) of the Legendre functions which breaks down for $\theta$ near zero or $\pi$.

In general, the cross-section (5.30) resembles a Fraunhofer diffraction pattern modified by the effects of the Coulomb repulsion, by the smooth cut-off in the absorption and by the refractive effects of the real nuclear potential. The Coulomb effects appear in a very simple way. They are contained in the Coulomb grazing angle $\theta_g$. The details of the absorptive and refractive effects of the optical potential are contained in the Fourier transform function $F(\chi)$ which is a factor in $J(\chi)$. As the phase $\eta(\theta)$ is a relatively slowly varying function of $\theta$ the spacing $\Delta\theta$ between peaks in the diffraction pattern is approximately

$$\Delta\theta \approx \pi/\lambda_g \tag{5.31}$$

irrespective of the strength of the Coulomb effects and the details of $F(\chi)$. This relation is very similar to the one (4.46) found in section 4.5 on the complex trajectory method.

Fig. 5.3 shows an example of an angular distribution for $^6Li + {}^{40}Ca$ elastic scattering at 156 MeV calculated from (5.31) with Ericson's parametrization. The parameter values are given in the figure caption. The diagram also shows the nearside and farside contributions to the cross-section calculated from (5.7) and (5.9) respectively. The nearside contribution dominates for small angles and the farside for large ones. The two contributions have equal magnitudes when $\theta = \bar{\theta} \simeq 19^\circ$ and the interference oscillations have maximum amplitude for $\theta$ near $\bar{\theta}$.

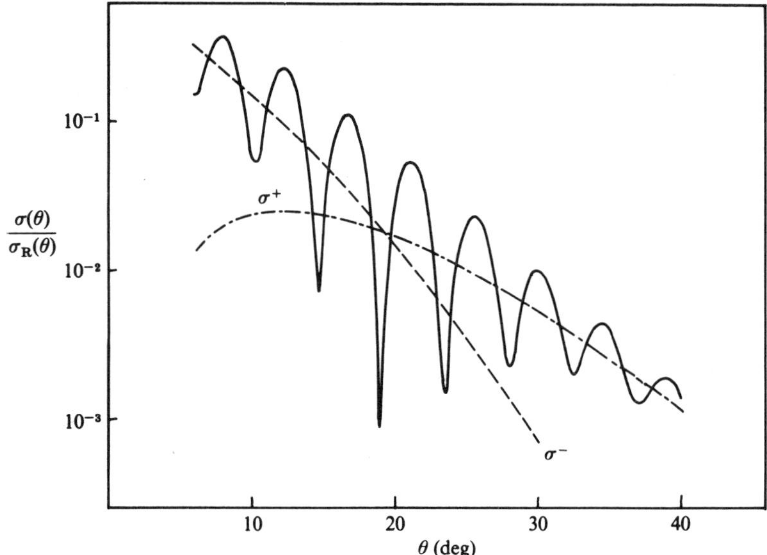


Fig. 5.3. Angular distribution for elastic scattering of $^6$Li by $^{40}$Ca at $E_{\text{lab}} =$ 156 MeV calculated with Ericson's parametrization (5.19). The parameters are $\lambda_g = 40$, $\alpha = .88$, $\Delta = 3.0$, $n = 1.86$, $\theta_g = 5.3°$.

There are simple approximate expressions for the ratio $b(\theta)$ for Ericson's and Kaufmann's parametrizations:
Ericson:

$$b(\theta) \approx \exp[2\Delta(\alpha\theta - \pi\theta_g)] \tag{5.32}$$

Kaufmann:

$$b(\theta) \approx \left(\frac{\theta_g - \theta}{\theta_g + \theta}\right)^{\frac{1}{2}} \exp[\Delta(2\tilde{\alpha}\theta - \pi\theta_g)] \tag{5.33}$$

If $b(\theta) < 1$ the nearside amplitude dominates the cross-section while for $b(\theta) > 1$ the farside part gives the larger contributions. A strong attractive (refractive) contribution from the nuclear interaction (large $\alpha$ or $\tilde{\alpha}$) favours the farside component and strong effects of Coulomb repulsion (large $\theta_g$) enhances the nearside contribution. Rowley & Marty (1976*a*) have discussed the nearside–farside interference in the case of Ericson's parametrization and have pointed out that $b(\theta) = 1$ when

$$\theta = \bar{\theta} = \pi\theta_g/\alpha \tag{5.34}$$

For this to happen in the physical region we need $\bar{\theta} < \pi$ or $\theta_g < \alpha$. When this condition is satisfied we can expect an exponentially decreasing cross-section due to the nearside amplitude $f_0^-(\theta)$ followed by an exponentially decreasing cross-section with a smaller slope due to the farside part $f_0^+(\theta)$. The angular spread $\delta\theta$ of the region of large oscillations is of order

$$\delta\theta \approx 1/(\Delta\alpha)$$

These qualitative aspects are demonstrated clearly in the example shown in fig. 5.3.

The condition $\theta_g < \alpha$ is necessary for the observation of nearside–farside interference, but is not sufficient. It can happen that the cross-section is so small for $\theta = \bar{\theta}$ given by (5.34) that it cannot be measured. A survey of experimental data shows that cross-sections with strong interference oscillations are characteristic of systems with a small Sommerfeld parameter. They are always observed for systems with $n < 4$ and have been seen only in one case ($^{16}O$, $^{28}Si$) with $n > 10$.

## 5.5 The choice of the grazing angular momentum

The method developed in this section is based on three stages of approximation:

(a) the Poisson sum formula (5.5),
(b) expansion (5.12) of the Coulomb phase,
(c) approximation (5.15) for $J(\chi)$.

The essential assumption in (b) and (c) is that the nuclear phase is a more rapidly varying function of $\lambda$ than the Coulomb phase. This is expressed in condition (5.4). Now we comment on the choice of $\lambda_g$. The theory is an approximation for any reasonable choice of $\lambda_g$, but a good choice should minimize the corrections.

If the scattering is of the Fresnel type then the theory predicts that the cross-section is one quarter of the Rutherford value at the grazing angle $\theta_g$ related to $\lambda_g$ by the Rutherford formula (2.2). A good choice of $\lambda_g$ would minimize the correction to this quarter-point property. Using equation (F.10) in appendix F the value of $J(\chi)$ at $\chi = 0$ $(\theta = \theta_g)$ is

$$J(0) = \tfrac{1}{2} - i(|\Theta_g'|/2\pi)^{\frac{1}{2}} e^{-\frac{1}{4}i\pi} F'(0)$$

The first term gives the quarter-point property and the second is a correction. The Fourier transforms calculated from parametrized amplitudes satisfy $|F'(0)| \sim \Delta$ and the magnitude of the correction is of order $(|\Theta_g'|\Delta^2/2\pi)^{\frac{1}{2}}$ which is small if the condition (5.4) holds. The best choice of $\lambda_g$ makes the correction term purely imaginary because then it contributes to the cross-section only in second order.

### 5.6 Coulomb rainbow versus Fresnel diffraction

The semi-classical methods of chapter 4 and the diffraction methods of chapter 5 have many common features. Both methods use the Poisson sum formula and agree that the amplitude is written as a sum of several components as indicated in fig. 4.7 or in fig. 5.4

$$\left.\begin{aligned} f(\theta) &\approx f_a^+(\theta) + f_a^-(\theta) + f_b^-(\theta) && \text{if}\,\theta < \theta_g \\ f(\theta) &\approx f_a^+(\theta) + f_d^-(\theta) && \text{if}\,\theta > \theta_g \end{aligned}\right\} \tag{5.35}$$

The two terms $f_a^-$ and $f_b^-$ for $\theta < \theta_g$ are the two components of the Coulomb rainbow in the semi-classical method discussed in section 4.4. The diffraction theory interprets them as a Rutherford amplitude and a diffracted amplitude. The region $\theta > \theta_g$ is a shadow region in both theories. In the rainbow theory it is a refractive shadow produced by the real part of the optical potential. The diffraction theory describes it as an absorptive shadow. In both theories $f_d^-$ is an amplitude which penetrates into the shadow region due to wave effects. The amplitude $f_a^+$ is a farside component which is bent around into the negative $\theta$ region by deflection in the semi-classical theory or diffraction in the Frahn theory. The amplitudes $f_a^-$ and $f_a^+$ have the same physical origin and $f_a^+$ is the continuation of the nearside component $f_a^-$ to the farside region. This is indicated by the dashed curves in fig. 5.4.

An important aspect of both theories is the asymmetry between the amplitudes $f_a^-$ and $f_d^-$. This is a property of the Fourier transform function $F(\theta_g - \theta)$ defined in equation (5.14) in the diffraction theory. If the nuclear amplitude $S_n(\lambda)$ is real i.e. purely absorptive then $F(\theta_g - \theta)$ is symmetric in $\theta_g - \theta$ and $f_a^-(\theta)$ and $f_d^-(\theta)$ fall off in the same way as $\theta$ moves away from $\theta_g$. An attractive real nuclear phase enhances $f_a^-$ and reduces $f_d^-$. The farside component $f_a^+$ is also enhanced because it is a continuation of $f_a^-$. There is an asymmetry between $f_a^-$ and $f_d^-$ from the start in the semi-classical theory because $f_a^-$ is associated with a classically allowed deflection while $f_d^-$ is in a classical shadow.

Each method has its own name for an angular distribution with the form shown in fig. 5.2. Proponents of the semi-classical method call it a Coulomb rainbow. A follower of Frahn's approach calls it Fresnel diffraction. The difference in name reflects the difference in view point. In the semi-classical method the Coulomb rainbow is a refractive effect and is produced when the attractive nuclear force balances the repulsive Coulomb force. The details of the shape of the rainbow are modified by the absorption represented by the imaginary part of the optical potential. In the diffraction model the Fresnel diffraction pattern is a consequence of the effects of strong

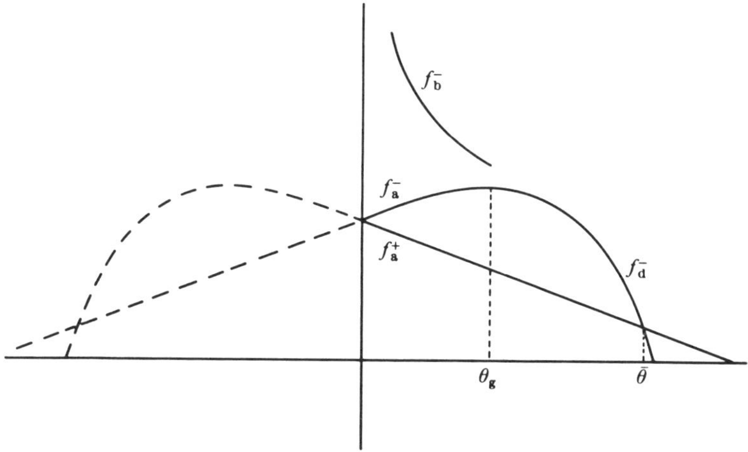


Fig. 5.4. Contributions of nearside amplitudes $f_a^-$, $f_b^-$ and $f_d^-$ and the farside amplitude $f_a^+$ to the cross-section (equation (5.35)).

absorption. The Blair quarter-point angle is determined mainly by the absorptive part of the $S$-matrix (the imaginary part of the phase shift). The phase of the $S$-matrix affects the details of the shape of the Fresnel pattern. It is difficult to decide which of the interpretations is the correct one because in many applications the real and the imaginary parts of the nuclear phase have comparable magnitudes. The complex trajectory method of Knoll & Schaeffer (1976) and Koeling & Malfliet (1975) adopts a neutral position because it treats the real and the imaginary parts of the nuclear potential on the same footing. At the end of this section we shall show how it is able to provide a link between the two approaches. But first we list some of the advantages and weaknesses of the two methods.

The diffraction model yields simple closed form expressions for the amplitudes $f_0$ and for the cross-sections in terms of the Fourier transform $F$ of the derivative of the nuclear partial wave scattering amplitude and there is a simple interpretation of angular distributions in terms of diffraction models. The Fresnel picture provides a natural explanation of the quarter-point property in terms of a grazing angular momentum $\lambda_g$ defined by $|S_n(\lambda_g)| \approx \frac{1}{2}$. Frahn (1984) has shown that the real part of the nuclear phase has only a minor effect on the quarter point property. The theory can be formulated in terms of a real angular momentum $\lambda$. Complex angular momenta and the saddle point method can be used to evaluate $F$ but this is an extra option and is not a necessity.

The diffraction theory is somewhat limited in its application and can only be used for a strong absorption amplitude as sketched in fig. 5.1. In fact, most applications have used a parametrized form for $S(\lambda)$ rather than one generated from an optical potential. It is relatively easy to impose simple physical constraints like unitarity on a parametrized $S$-matrix, but such a form can have other unphysical properties which are not easy to spot. For example, the Frahn–Venter parametrization can produce an unphysical anomaly in a Fresnel angular distribution. This happens when the product $\alpha\Delta$ is large and the factor $(1+\alpha\Delta(\theta_g-\theta))$ in (5.22*a*) is zero for some value of $\theta<\pi$.

The stationary phase approach to the semi-classical method also has some very nice features. It is possible to generate a rainbow angular distribution (fig. 4.5) using the Airy uniform approximation directly from a deflection function like fig. 4.2 which shows a Coulomb rainbow. The method treats absorption as a perturbation which may not be very accurate for some applications. The complex trajectory method does not suffer from this defect but calculations with complex saddle points and complex angular momenta are cumbersome.

We conclude this section by showing that Frahn's method and the complex trajectory method are equivalent provided

(a) The strong absorption condition (5.4) of Frahn's theory holds.
(b) The saddle point method for evaluating the integrals $f_0^{\pm}(\theta)$ is accurate.

The diffraction model is formulated in terms of the Fourier transform function $F(\chi)$ (5.14). Integrating by parts it can also be written as

$$F(\chi)=-i\chi\int_0^\infty d\lambda S_n(\lambda)\exp\{i\chi(\lambda-\lambda_g)\} \tag{5.36}$$

The complex trajectory method is based on a saddle point approximation to integrals of the form

$$I(\theta)=\int_0^\infty \lambda^{\frac{1}{2}}d\lambda S_n(\lambda)\exp(2i\sigma(\lambda)\pm i\lambda\theta) \tag{5.37}$$

When $S_n(\lambda)$ is a strong absorption amplitude then the saddle points $\lambda_s$ associated with the components $f_a^{\pm}(\theta)$ and $f_d^{-}(\theta)$ are both close to the grazing angular momentum $|\lambda_s-\lambda_g|\lesssim\Delta$. If condition (5.4) for the validity of the diffraction model holds then $\lambda^{\frac{1}{2}}$ in (5.37) can be approximated by $\lambda_g^{\frac{1}{2}}$ and the Coulomb phase can be expanded as

$$2\sigma(\lambda)\simeq 2\sigma(\lambda_g)+\theta_g(\lambda-\lambda_g) \tag{5.38}$$

neglecting quadratic terms. With these approximations the integral $I(\theta)$ is proportional to $F(\chi)$, the saddle point evaluation of $I(\theta)$ is almost the same as the saddle point approximation for $F(\chi)$ and the two methods are equivalent. This argument holds only if $\theta$ is not too close to the grazing angle $\theta_g$. When $|\theta_g - \theta|\Delta \lesssim 1$ the saddle point evaluations of $F(\chi)$ and $I(\theta)$ are not valid and the expansion (5.38) is not accurate enough. In this region the methods are different. The diffraction method gives an interpolation based on Fresnel integrals while the complex trajectory method uses an Airy uniform approximation.

## 5.7 Large angle scattering

There is a factor of $(\sin\theta)^{-1}$ in equation (4.1) for the classical cross-section. Hence the cross-section becomes infinite for $\theta = \pi$ if back-scattering is possible for impact parameters other than zero. This is the analogue of the optical glory effect in scattering of light by water droplets (Nussenzweig, 1979). The $(\sin\theta)^{-1}$ singularity is due to axial symmetry of the scattering and Nussenzweig calls it the axial focusing effect. Axial focusing also occurs at $\theta = 0$, but in heavy-ion elastic scattering this forward focusing is hidden by the Rutherford scattering. The glory singularity is spread out by wave effects in a semi-classical theory and produces a characteristic cross-section for large angle scattering. Studies by Nussenzweig (1979) and others have shown that the optical glory for scattering by a refracting sphere is very complicated and the amplitude has contributions from waves which suffer many internal reflections before emerging from the scatterer. The heavy-ion glory effect is simpler because the absorption is relatively strong and components associated with multiple reflections are damped out. In most cases the dominant contribution comes from surface waves and we shall study this part in the present section. There can be contributions from internal waves for some light systems (eg. ($\alpha$, $^{40}$Ca), ($^{16}$O, $^{28}$Si)). These will be discussed in chapter 6.

The components of the scattering amplitude (5.35) which are illustrated in fig. 5.4 all come from the $m = 0$ term in the Poisson sum formula (4.16), but if the scattering angle is near $\pi$ other terms have to be considered. Fig. 5.5 illustrates two cases where this happens. The observation angle is $\theta = \pi - \vartheta$ for all the trajectories illustrated while the deflection angle is $-\pi + \vartheta$ in fig. 5.5(*a*) and $-\pi - \vartheta$ in fig. 5.5(*b*). If the deviation $\vartheta$ of the angle of observation from $\pi$ is small then it is clear that both contributions are important, since the amplitude $f_1^-(\pi - \vartheta)$ is the extension of $f_0^+(\pi - \vartheta)$ to negative angles and both terms have comparable magnitudes. The Poisson amplitudes associated with fig. 5.5(*c*) and 5.5(*d*) are $f_0^-(\pi - \vartheta)$ and $f_{-1}^+(\pi - \vartheta)$ respectively and the corresponding deflection

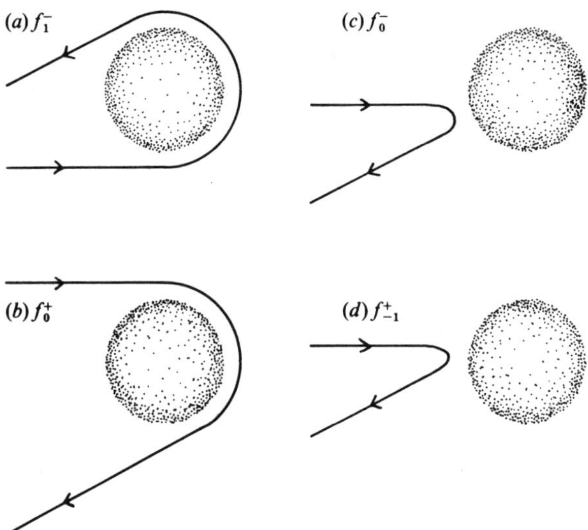


Fig. 5.5. Illustration of the classical orbits corresponding to several terms of the Poisson sum formula which contribute at backward angles. (*a*) $f_1^-: \Theta = -\pi - \vartheta$, (*b*) $f_0^+: \Theta = -\pi + \vartheta$, (*c*) $f_0^-: \Theta = \pi - \vartheta$, (*d*) $f_{-1}^+: \Theta = \pi + \vartheta$.

angles are $\pi - \vartheta$ and $\pi + \vartheta$. The second of these is a forbidden angle for classical scattering, but there can be a diffractive quantal contribution for these classically forbidden angles. The basic results for large angle scattering are already contained in the work of Ford & Wheeler (1959), Bryant & Jarmie (1968) and Berry (1969). The theory has been applied to large angle heavy-ion scattering by Takigawa & Lee (1977), Frahn (1980*a*) and others.

We first consider the cases illustrated in fig. 5.5(*a*, *b*) when the classical deflection angle is near $-\pi$. Bryant & Jarmie have given a rough formula for the back angle cross-section in terms of a Bessel function

$$\sigma(\theta) \approx AJ_0^2(\lambda_g \vartheta) \tag{5.39}$$

where $\theta = \pi - \vartheta$ and $\lambda_g$ is a grazing angular momentum. Fig. 5.6 shows some examples of $\alpha$-particle elastic back scattering from spinless nuclei compared with equation (5.39) which were given by Bryant & Jarmie. The fits in all cases down to the region of 150° are generally good. The minima in the cross-section occur at the zeros $u_1 = 2.40$, $u_2 = 5.52$, etc., of the Bessel function $J_0(u)$. The minima are relatively well defined in the experimental data and the value of the parameter $\lambda_g$ can be extracted from their positions.

Other theories give generalizations of the simple formula (5.39). In Frahn's diffraction theory the term $f_0^+(\theta)$ is given by (5.9) and $f_1^-(\theta)$ can

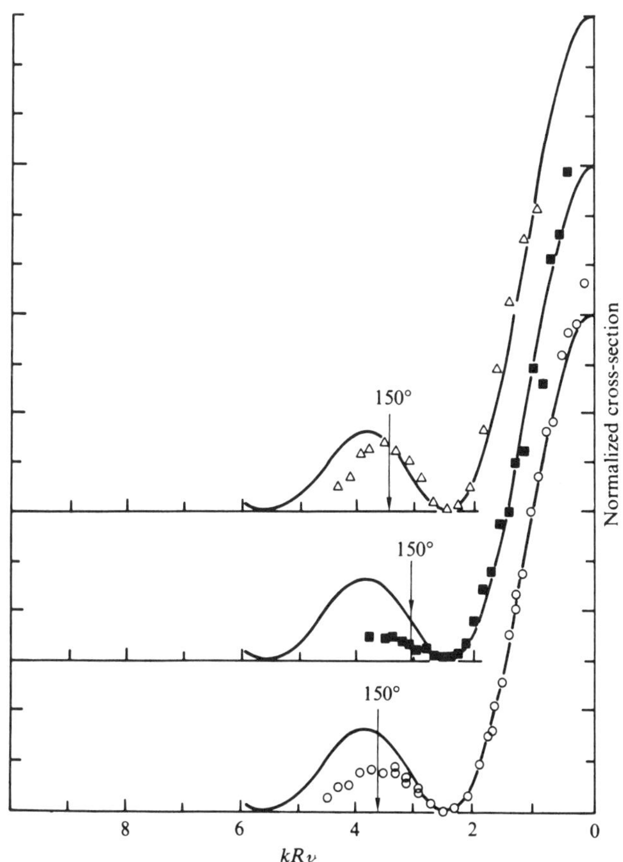


Fig. 5.6. Examples of α-particle back scattering with fits from the nuclear glory model (equation 5.39) (from Bryant & Jarmie, 1968), (*a*) from $^{16}O$ at 18.3 MeV, (*b*) from $^{12}C$ at 21.9 MeV, (*c*) from $^{12}C$ at 18.5 MeV.

be obtained by a similar argument. There is an additional phase factor $\exp(-\frac{1}{2}i\pi)$ in $f_1^-$ which comes from the phases in the Poisson formula (4.16) and is associated with the axial focusing at $\theta = \pi$ (see section 4.6). The result written in a way which shows the relation between the $f_0^+$ and $f_1^-$ contributions is

$$\begin{aligned} f(\theta) &\approx f_0^+(\pi - \vartheta) + f_1^-(\pi - \vartheta) \\ &= (\sin\theta)^{-\frac{1}{2}}[\mathscr{R}(-\vartheta)\exp(-i\lambda_g\vartheta + \tfrac{1}{4}i\pi) + \mathscr{R}(\vartheta)\exp(i\lambda_g\vartheta - \tfrac{1}{4}i\pi)] \end{aligned} \tag{5.40}$$

$$\mathscr{R}(\vartheta) = f_C(\theta_g)(\sin\theta_g)^{\frac{1}{2}}J(\theta_g + \pi + \vartheta)\exp i(\lambda_g(\theta_g + \pi) + \tfrac{1}{4}\pi) \tag{5.41}$$

Equation (5.40) gives a cross-section

$$\left.\begin{array}{l}\sigma(\theta)=|f_0^+(\pi-\vartheta)|^2[1+r^2(\vartheta)+2r(\vartheta)\cos(2\lambda_g\vartheta)+\eta(\vartheta)]\\ r(\vartheta)=|\mathscr{R}(-\vartheta)/\mathscr{R}(\vartheta)|,\quad \eta(\vartheta)=\arg(\mathscr{R}(\vartheta)/\mathscr{R}(-\vartheta))\end{array}\right\} \tag{5.42}$$

In (5.42) $|f_0^+(\pi-\vartheta)|^2$ is the cross-section calculated from the farside amplitude (5.9). With Ericson's parametrization (5.19) the phase $\eta(\vartheta)\simeq 0$ and the ratio

$$r(\vartheta)=\exp(-2\Delta(\pi-\alpha)\vartheta) \tag{5.43}$$

The formula (5.42) predicts oscillations in the cross-section due to farside–nearside interference with spacing

$$\Delta\theta\approx\pi/\lambda_g \tag{5.44}$$

These have a maximum amplitude for large angle scattering and die away for smaller angles because the ratio $r(\vartheta)$ tends to zero as $\vartheta$ increases. The cross-section (5.42) contains a factor $1/\sin\theta$ which diverges when $\theta=\pi$. This is the same divergence as in the classical cross-section (4.1) and occurs in (5.42) because one of the approximations made in deriving it is not accurate for $\theta\simeq\pi$. The asymptotic formula (4.14) for the Legendre polynomial breaks down when $\vartheta=\pi-\theta\lesssim l^{-1}$ and the Bessel function approximation (4.19) should be used instead. Then the Poisson formula (4.15) is replaced by

$$f(\theta)=\frac{1}{k}\left(\frac{\vartheta}{\sin\vartheta}\right)^{\frac{1}{2}}\sum_n(-1)^n\int_0^\infty \lambda d\lambda S(\lambda)J_0(\lambda\vartheta)\exp((2n-1)i\pi\lambda) \tag{5.45}$$

When $\lambda\vartheta\gg 1$ the Bessel function in (5.45) can be replaced by its asymptotic form (A.15) and (5.45) reduces to the old form (4.16) of the Poisson expansion. However, the terms are collected in a different way. The trajectories in fig. 5.5($a,b$) are contained in the term $n=1$ while the trajectories in fig. 5.5($c,d$) correspond to $\mathrm{n}=0$.

A modification of (5.40) which is valid even when $\vartheta=0$ can be derived by evaluating the $n=1$ integral in (5.45) by a uniform approximation (appendix C). The generalized equation is

$$\begin{aligned}f(\theta)=\left(\frac{\pi\lambda_g\vartheta}{2\sin\vartheta}\right)^{\frac{1}{2}}&[\mathscr{R}(-\vartheta)(J_0(\lambda_g\vartheta)-iJ_1(\lambda_g\vartheta))\\ &+\mathscr{R}(\vartheta)(J_0(\lambda_g(\vartheta)+iJ_1(\lambda_g\vartheta))]\end{aligned} \tag{5.46}$$

The asymptotic formula (A.17) for the Bessel function

$$J_0(z)\mp iJ_1(z)\sim(2/\pi z)^{\frac{1}{2}}\exp[\mp(z-\tfrac{1}{4}\pi)]$$

which holds for large $z$, ensures that (5.46) reduces to (5.40) when $\lambda_g\vartheta \gg 1$. The amplitude when $\theta = \pi (\vartheta = 0)$ is finite and

$$f(\pi) = (2\pi)^{\frac{1}{2}} \lambda_g \mathscr{R}(0) \tag{5.47}$$

The uniform approximation (5.46) for large angle scattering was obtained starting from Frahn's diffraction theory. The stationary phase and complex trajectory methods developed in chapter 4 yield results with much the same general structure as (5.46). The difference lies mainly in the expression for $\mathscr{R}(\vartheta)$ which depends on the classical deflection function in the stationary phase method. There is also a minor difference in the argument of the Bessel function which now depends on the stationary angular momentum $\lambda_s(\theta)$ rather than the grazing angular momentum in Frahn's approach. The stationary phase analogue of (5.45) has been used by Takigawa & Lee (1977) to analyse large angle ($\alpha$, $^{40}$Ca) scattering.

Ericson's partial wave amplitude (5.19) has a line of poles in the complex $\lambda$-plane at $\lambda = \lambda_g + ((2r-1)\pi - \alpha)\Delta i$ where $r$ is an integer. We used this fact in section 5.2 in order to obtain an approximate evaluation (5.23) of the Fourier transform function $F(\chi)$ which appears in Frahn's formulation of semi-classical heavy-ion scattering. The same method can be used to make a simple evaluation of the integrals in the Poisson formula (5.45) for the large angle amplitude. The integral in the $n = 1$ Poisson term can be calculated by closing the contour for the $\lambda$-integration in the upper half $\lambda$-plane. Provided $\Delta$ is not too small ($\pi\Delta \gtrsim 1$) the dominant contribution comes from the nearest pole in the upper half-plane at

$$\lambda_1 = \lambda_g + (\pi - \alpha)\Delta i \tag{5.48}$$

Evaluating the residue of the pole gives

$$f(\theta) \simeq -\left[\frac{2\pi i \Delta \lambda_1}{k}\right]\left(\frac{\vartheta}{\sin\vartheta}\right)^{\frac{1}{2}} J_0(\lambda_1\vartheta)\exp[i(2\sigma(\lambda_1) + \pi\lambda_1)] \tag{5.49}$$

Although (5.46) and (5.49) look rather different, they are equivalent when (5.46) is evaluated using Ericson's amplitude and when $\lambda_g$ is large ($\lambda_g \gg \pi\Delta$). Equation (5.49) was obtained by Takigawa & Lee (1977) and is interesting because it gives a very simple generalization of (5.39) for the large angle amplitude. When $\theta = \pi$ then $\vartheta = 0$ and the Bessel function in (5.49) is unity

$$f(\pi) = -\left[\frac{2\pi i \Delta \lambda_1}{k}\right]\exp[i(2\sigma(\lambda_1) + \pi\lambda_1)] \tag{5.50}$$

Expanding the Coulomb phase about $\lambda_g$ to first order

$$2\sigma(\lambda_1) \simeq 2\sigma(\lambda_g) + i\theta_g(\pi - \alpha)\Delta$$

and using equation (2.4) for the Rutherford cross-section we obtain

$$\frac{\sigma(\pi)}{\sigma_{\mathrm{R}}(\pi)} \simeq \left(\frac{4\pi\Delta\lambda_{\mathrm{g}}}{n}\right)^2 \exp\left[-2(\theta_{\mathrm{g}}+\pi)(\pi-\alpha)\Delta\right] \tag{5.51}$$

The contributions of the orbits illustrated in figs. 5.5(*c*, *d*) are important for scattering at energies near or below the Coulomb barrier. Landowne & Wolter (1981) have developed a theory of sub-Coulomb elastic scattering and we present some of their results here. Low angular momenta give the dominant contribution to back angle scattering at sub-Coulomb energies. The nuclear phase shifts are small in magnitude and vary smoothly with the semi-classical angular momentum $\lambda$. Landowne & Wolter use the stationary phase formula

$$f(\theta) = \left[\frac{\lambda}{k^2 \sin\theta |\Theta'(\lambda)|}\right]^{\frac{1}{2}} \exp i(2\delta(\lambda) - \lambda\theta - \tfrac{1}{2}\pi) \tag{5.52}$$

where $\Theta'(\lambda) = 2\,\mathrm{Re}\,\delta''(\lambda)$. This follows from equations (4.24) and (4.33) if we assume that $f(\theta) \simeq f_0^-(\theta)$.

At first sight it appears that (5.52) cannot be used for large angle scattering because of the following reasons:

(a) The asymptotic form (4.14) for the Legendre polynomials used to derive equation (4.24) fails when $\theta \simeq \pi$.
(b) The stationary phase formula (4.24) is invalid since the stationary angular momentum $\lambda(\theta)$ is near the end point $\lambda = 0$ of the integral (4.20).
(c) The farside contribution $f_{-1}^{+}(\theta)$ illustrated in fig. 5.5(*d*) has been neglected in (5.52).

In fact, the errors due to these defects in the theory cancel and equation (5.52) is a good semi-classical formula for $f(\theta)$, even when $\theta = \pi$. This can be shown by deriving (5.52) in a way which avoids the problems mentioned above. We start with the $n = 0$ term in equation (5.45). This includes the amplitudes $f_0^-(\theta)$ and $f_1^+(\theta)$ and should be accurate for $\theta = \pi$ because it uses the Bessel function asymptotic form (4.19) for the Legendre polynomials. The next step is to introduce a two-dimensional angular momentum vector $\boldsymbol{\lambda} = k\mathbf{b}$ where $\mathbf{b}$ is the impact parameter vector and use the integral representation (A.10) for the Bessel function. Then the $n = 0$ term in (5.45) can be written as

$$f(\theta) = \frac{1}{2\pi k}\left(\frac{\vartheta}{\sin\theta}\right)^{\frac{1}{2}} \int_{-\infty}^{\infty}\int_{-\infty}^{\infty} d\lambda_x d\lambda_y \exp i(\lambda_x \vartheta + 2\delta(\lambda) - \lambda\pi) \tag{5.53}$$

When $\lambda$ is small the Coulomb phase is

$$\sigma(\lambda) \simeq \sigma_0 + \lambda\pi - \lambda^2/n \tag{5.54}$$

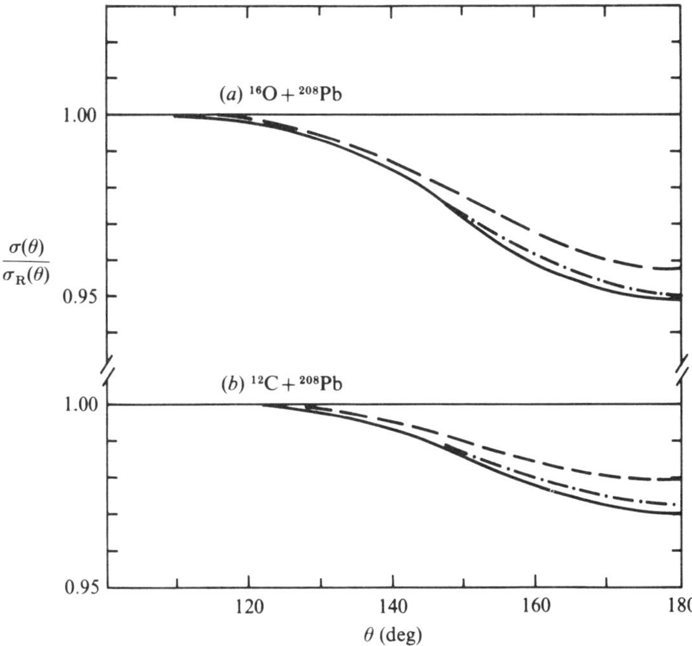


Fig. 5.7. Examples of sub-Coulomb elastic scattering, (*a*) $^{16}O + ^{208}Pb$ at $E_{lab} =$ 72.95 MeV, (*b*) $^{12}C + ^{208}Pb$ at $E_{lab} = 54.07$ MeV. The solid curve is calculated from an optical model, the dashed curve from equation (5.57) and the dot–dashed curve from Traber *et al.* (1977) (from Landowne & Wolter, 1981).

If the approximation (3.66) is used for the nuclear phase then for small $\lambda$

$$\delta_n(\lambda) \simeq \delta_n(0) \exp[-\lambda^2/(2nka)] \tag{5.55}$$

Hence $2\delta(\lambda) - \lambda\pi$ is an even function of $\lambda$ which is regular at $\lambda = 0$ and (5.53) can be evaluated by the stationary phase approximation in two dimensions (appendix B). There is no difficulty with (a) because the Bessel approximation (4.19) is used, with (b) because the integration limits are $\pm\infty$ or with (c) because (5.53) includes the contribution of $f_0^-$ and $f_{-1}^+$. The integral (5.53) has one stationary point and evaluating it by the method described in appendix B leads to the amplitude (5.52).

Equation (5.52) for the amplitude $f(\theta)$ yields the classical cross-section (4.34).

$$\sigma(\theta) = \left[\frac{\lambda}{k^2 \sin\theta |\Theta'(\lambda)|}\right] \exp(-4 \operatorname{Im} \delta_n(\lambda)) \tag{5.56}$$

where $\lambda$ is the angular momentum corresponding to the classical deflection angle $\theta = \pi - \vartheta$. Landowne & Wolter (1981) have given a simple expression

Table 5.1 *Potential parameters for sub-Coulomb scattering (fig. 5.1)*

| | $V_0$ | $r_v$ | $a_v$ | $W_0$ | $r_w$ | $a_w$ |
|---|---|---|---|---|---|---|
| $^{12}C + ^{208}Pb$ | 30.0 | 1.31 | 0.5 | 5.0 | 1.31 | 0.5 |
| $^{16}O + ^{208}Pb$ | 30.0 | 1.31 | 0.5 | 10.0 | 1.31 | 0.5 |

$V_0$ and $W_0$ in MeV, $r_v$, $r_w$, $a_v$ and $a_w$ in fm.

for (5.56) in the limiting case where $|\delta_n(\lambda)| \ll 1$.

$$\sigma(\theta)/\sigma_R(\theta) = 1 - \frac{(2\pi nka)^{\frac{1}{2}}}{E}\left[2W(r_C) + \frac{V(r_C)}{ka}\left(1 - \frac{\lambda^2}{2nka}\right)\right] \tag{5.57}$$

Equation (5.57) uses the phase shift formula (3.66), and $r_C$ is the distance of closest approach (3.67) in a Rutherford orbit with angular momentum $\lambda$ and scattering angle $\theta$. Fig. 5.7 shows an example of the way in which the Rutherford cross-section is modified by the nuclear optical potential for sub-Coulomb scattering. The calculations are made using a Woods–Saxon potential with parameters given in table 5.1. The cross-section is reduced below the Rutherford value because of absorption by the imaginary part of the optical potential and because the attractive real potential deflects some of the back-angle cross-section towards more forward angles. Equation (5.57) shows that the cross-section is more sensitive to the imaginary potential than the real potential for large *ka*. For the cases illustrated in fig. 5.7 ($ka \approx 3$).

The semi-classical result (5.56) is very similar to a more rigorous one obtained by Traber *et al.* (1977) who applied quantum mechanical perturbation theory to a given optical potential. This is interesting because (5.56) is essentially a classical formula. The similarity suggests that quantal effects are not very important for sub-Coulomb scattering.

# 6

# Surface transparent potentials

## 6.1 The three-turning point problem

The discussion of elastic scattering in chapters 3, 4 and 5 was restricted to the case of a strongly absorbing optical potential. The scattering was sensitive only to the potential outside the Coulomb barrier and the semi-classical phase shift (3.49) had a contribution only from the outermost turning point in the barrier region. In this chapter we consider optical potentials with weaker absorption so that the scattering is sensitive to the potential at and inside the Coulomb barrier. The aim is to study the new effects which appear in such cases.

In the present section we study the structure of the partial wave scattering amplitude when the effective potential has three classical turning points. A typical case is shown in fig. 6.1. There is a classically allowed region between the turning points $r_3$ and $r_2$ and for $r > r_1$ and a classically inaccessible region between $r_2$ and $r_1$. We assume that the turning point $r_3$ is well separated from $r_2$ and $r_1$ but that $r_2$ and $r_1$ may be close together. If the optical potential has an imaginary part then the turning points will be complex as discussed in section 3.4. Their positions in the complex $r$-plane are indicated in the lower part of fig. 6.1.

If the incident energy is near the top of the barrier all three turning points should be taken into account. We shall show that the resulting amplitude is (Brink & Takigawa, 1977)

$$S_n(\lambda) \approx \exp(2i\delta_1(\lambda)) \frac{1 + \bar{N}(i\beta)\exp(2iW_{32})}{N(i\beta) + \exp(2iW_{32})} \tag{6.1}$$

In equation (6.1) $\delta_1(\lambda)$ is the nuclear phase calculated with respect to the outer turning point $r_1$. It is given by formula (3.49) in section 3.8. The remaining factor contains the effects of barrier penetration and the internal region on $S_n(\lambda)$. In the following we use the notation

$$w(r, r') = \int_{r'}^{r} k(r)\, dr \tag{6.2}$$

with the local wave number $k(r)$ given by equation (3.50). The quantities

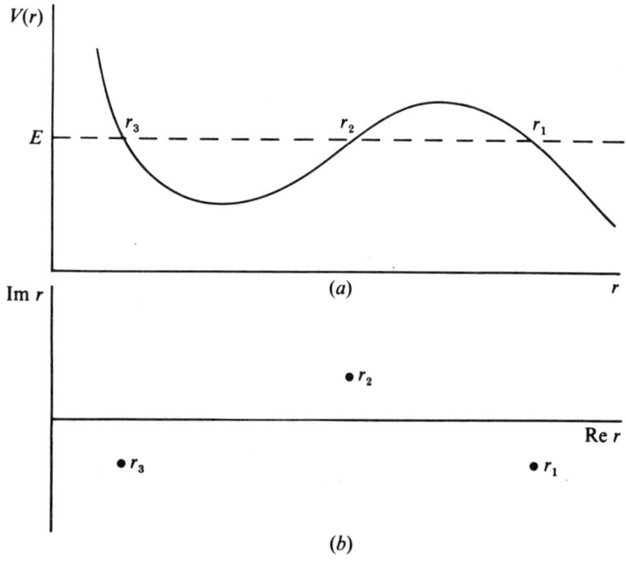


Fig. 6.1. (*a*) Sketch of a potential $V(r)$ with three turning points ($V(r) = E$) at $r = r_1$, $r_2$, $r_3$. (*b*) Positions of turning points in the complex $r$-plane if $V(r)$ is complex and Im $V(r) < 0$.

$\beta$ and $W_{32}$ in (6.1) are defined by

$$i\pi\beta = w(r_1, r_2); \quad W_{32} = w(r_3, r_2) \tag{6.3}$$

The turning points are complex and the integrals (6.3) must be taken along suitable paths in the complex $r$-plane. The local wave number $k(r)$ defined by (3.50) has branch points at the turning points $r_1$, $r_2$ and $r_3$. The branch cuts are chosen so that $k(r)$ is an analytic function on and near the real axis. The integrals for $\beta$ and $W_{32}$ are independent of the integration paths provided that these lie on the first Riemann sheet. The functions $N(i\beta)$ and $\bar{N}(i\beta)$ are given by equations (E.12) and (E.13) in appendix E. They depend on the action integral (6.3) under the barrier and take into account barrier penetration effects.

We now outline the derivation of (6.1) (Lee and Takigawa 1978). In the region $r > r_1$, the WKB wave function has the form (3.4)

$$\psi(r) = (k(r))^{-\frac{1}{2}}(C_+ \exp(iw(r, r_1)) + C_- \exp(-iw(r, r_1)) \tag{6.4}$$

with $w(r, r_1)$ given by (6.2). The first step is to find a relation between $C_+$ and $C_-$. In the interior region $r_3 < r < r_2$ the WKB wave function can be written as

$$\psi(r) = (k(r))^{-\frac{1}{2}}(B_+ \exp(iw(r, r_2)) + B_- \exp(-iw(r, r_2))) \tag{6.5}$$

The coefficients $C_{\pm}$ are related to $B_{\pm}$ by the barrier matrix $F$ (appendix E)

$$\begin{pmatrix} C_+ \\ C_- \end{pmatrix} = \begin{pmatrix} f_{11} & -if_{12} \\ if_{21} & f_{22} \end{pmatrix} \begin{pmatrix} B_+ \\ B_- \end{pmatrix} \tag{6.6}$$

$$f_{11} = e^{\pi\beta}\bar{N}(i\beta), \quad f_{22} = e^{\pi\beta}N(i\beta)$$

$$f_{12} = f_{21} = e^{\pi\beta}$$

The wave function (6.5) in $r_3 < r < r_2$ can also be written with respect to the turning point at $x = r_3$ as

$$\psi(r) = (k(r))^{-\frac{1}{2}}(A_+ \exp(iw(r, r_3)) + A_- \exp(-iw(r, r_3)))$$

where

$$A_{\pm} = B_{\pm} \exp(\mp iW_{32})$$

where $W_{32}$ is given by (6.3). Now the matching condition (3.26) in section (3.3) gives $A_+ = -iA_-$ so that

$$B_+ = -iB_- \exp(2iW_{32})$$

Substituting into the connection formula (6.6) yields

$$\frac{C_+}{C_-} = -i\frac{1 + \bar{N}(i\beta)\exp(2iW_{32})}{N(i\beta) + \exp(2iW_{32})} \tag{6.7}$$

The connection formula (3.26) at an isolated turning point would give $(C_+/C_-) = -i$. The additional factor in (6.7) gives the effect of barrier penetration and the internal region. The semi-classical formula for the nuclear scattering amplitude at an isolated turning point $r_1$ would be $\exp(2i\delta_1)$. The same factor in (6.7) gives the modification due to barrier penetration and the internal region and leads to the result stated in (6.1). Formulae equivalent to (6.1) for a real potential have been given by several authors (see Berry & Mount 1972). The uniform approach of appendix E was used by Conner (1973). The particular form (6.1) is well adapted for generalizing to complex potentials and was used by Brink & Takigawa (1977) to study $\alpha$-scattering.

If there is a quasi-bound state inside the potential barrier then there is a resonance in the reflected amplitude $C_+$. For a real potential $V(r)$ the results in appendix E show that

$$N = (1 + e^{-2\pi\beta})^{\frac{1}{2}}e^{i\phi}, \quad \bar{N} = N^*$$

There is a resonance when the denominator of (6.7) or (6.1) is almost zero, that is, when $\exp(2iW_{32} - i\phi) = -1$. This gives

$$W_{32} = \int_{r_3}^{r_2} k(r)\,dr = (n + \tfrac{1}{2})\pi + \tfrac{1}{2}\phi \tag{6.8}$$

This is just the Bohr–Sommerfeld quantization condition with a small shift due to barrier penetration coming from the phase of $N$. Near the resonance energy

$$\begin{aligned}\exp(2iW_{32}(E)-i\phi) &\approx -1-2i(E-E_r)(\partial W_{32}/\partial E)_{E_r} \\ &= -1-i(E-E_r)T/\hbar\end{aligned} \tag{6.9}$$

In equation (6.9) the energy variation of $\phi$ has been neglected while the energy derivative of $W_{32}$ has been related to the period $T$ of the classical motion in the potential pocket inside the barrier (fig. 6.1). Substituting in (6.7) and assuming that the barrier penetration factor is small

$$P=(1+e^{2\pi\beta})^{-1}\approx e^{-2\pi\beta}\ll 1$$

one obtains a resonance formula

$$S_n(\lambda)\approx\exp(2i\delta_1(\lambda))\left(\frac{E-E_r-\frac{1}{2}i\Gamma_0}{E-E_r+\frac{1}{2}i\Gamma_0}\right) \tag{6.10}$$

with width $\Gamma_0=\hbar P/T$. The width $\Gamma_0$ of the resonance (6.10) is due to the decay of the compound state by barrier penetration. If the potential $V(r)$ is complex then there is an additional width due to absorption. When the absorption is not too big (6.10) is replaced to a rough approximation by

$$S_n(\lambda)\approx\exp(2i\delta_1(\lambda))\left[1-\frac{i\Gamma_0}{E-E_r+\frac{1}{2}i\Gamma}\right]$$

with $\Gamma\approx\Gamma_0+2\bar{W}$ and $\bar{W}$ is the average imaginary potential inside the barrier.

## 6.2 Regge poles and diffracted trajectories

The methods developed in chapters 4 and 5 lead to a decomposition of the amplitude

$$f(\theta)\approx f_a^+(\theta)+f_a^-(\theta)+f_b^-(\theta);\quad \theta<\theta_g \tag{5.34}$$

$$f(\theta)\approx f_a^+(\theta)+f_a^-(\theta);\qquad \theta>\theta_g \tag{5.35}$$

which is illustrated in fig. 5.4 in section 5.6. When $\theta<\theta_g$ there is an 'outer' or Coulomb contribution $f_b^-$ and an 'inner' or nuclear contribution $f_a^\pm$ which has nearside and farside components. The semi-classical method describes $f_a^\pm$ as due to the refractive Coulomb rainbow effect, that is, it is produced by trajectories deflected and damped by the nuclear attraction and absorption. Fuller & Moffa (1977) assert that there are cases where this is not the correct description, but that the inner contribution $f_a^\pm$ is due to a Regge pole in the nuclear partial wave amplitude. This contribution can be understood in terms of a generalized trajectory which decays as it

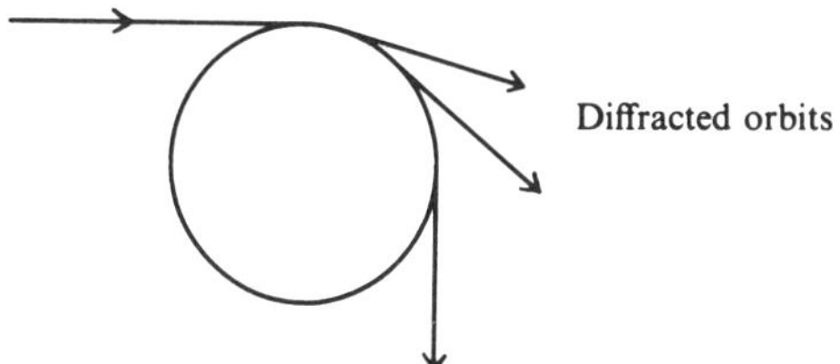


Fig. 6.2. Illustration of 'diffracted trajectories' associated with Regge poles.

propagates around the interaction surface. Such trajectories were introduced into diffractive theory by Keller (1958) and Levy & Keller (1959). They are known there as 'diffracted rays'. Fuller & Moffa (1977) develop the analogous concept of 'diffracted trajectories' for nuclear scattering problems. These diffracted trajectories are illustrated in fig. 6.2. We begin this section by using Frahn's method to show that the 'diffracted trajectories' associated with Regge poles contribute to the farside amplitude $f_0^+(\theta)$ and to the nearside amplitude $f_0^-(\theta)$ when $\theta < \theta_g$, but give no contribution to the nearside amplitude $f_d^-(\theta)$ in the shadow region $\theta > \theta_g$. Then we evaluate the pole contribution directly from the integral (4.17), we discuss the pole contribution to large angle scattering and finally make some comments on the method of Fuller & Moffa.

When using Frahn's method for calculating an elastic scattering amplitude it is necessary to evaluate the Fourier transform $F(\chi)$ defined in equation (5.14). Integrating (5.14) by parts gives

$$F(\chi) = -i\chi \int_0^\infty d\lambda S_n(\lambda) \exp(i\chi(\lambda - \lambda_g)) \tag{6.11}$$

with $\chi = \theta_g - \theta$. In certain cases the dominant contributions to $F(\chi)$ come from Regge poles. A scattering amplitude $S_n(\lambda)$ can have poles in the first quadrant of the complex $\lambda$-plane ($\mathrm{Re}\,\lambda > 0$, $\mathrm{Im}\,\lambda > 0$) but not in the fourth quadrant ($\mathrm{Re}\,\lambda > 0$, $\mathrm{Im}\,\lambda < 0$). When $\theta_g > \theta$ or $\chi > 0$ the exponential factor in (6.11) decreases in magnitude if $\mathrm{Im}(\lambda)$ increases. Then the integration contour in (6.11) can be moved into the upper part of the complex $\lambda$-plane (fig. 6.3) so that

$$F(\chi) = \sum_\alpha (-2\pi\chi\beta_\alpha) \exp\{i\chi(\lambda_\alpha - \lambda_g)\} - i\chi \int_C d\lambda S_n(\lambda) \exp\{i\chi(\lambda - \lambda_g)\} \tag{6.12}$$

Here $\lambda_\alpha$ is the position of a pole of $S_n(\lambda)$ and $\beta_\alpha$ is the corresponding residue. The summation in (6.12) is over all poles between the real $\lambda$-axis and the

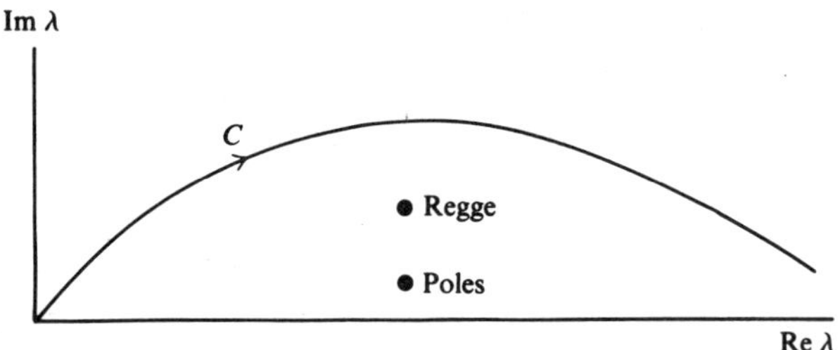


Fig. 6.3. Integration contour $C$ for calculating the contributions of Regge poles to the scattering amplitude.

contour $C$. This modification of the integration contour is particularly useful if there is only one isolated Regge pole $\lambda_1$ near the real axis and if the background integral over the contour $C$ is negligible. When $\chi$ is large enough the dominant contribution to $F(\chi)$ comes from the pole at $\lambda_1$ nearest to the real axis and

$$F(\chi) \approx (-2\pi\chi\beta_1)\exp\{i\chi(\lambda_1 - \lambda_g)\} \tag{6.13}$$

This gives a contribution to the nearside amplitude (5.8) and the farside amplitude (5.9) which decreases exponentially as $\theta$ moves away from the grazing angle.

When $\theta_g < \theta$ or $\chi < 0$ it is not useful to deform the integration contour into the upper part of the $\lambda$-plane as in fig. 6.3 because then the exponential factor in (6.12) increases with $\operatorname{Im}\lambda$, the integral along the background contour $C$ is large and it cannot be neglected. It might be useful to move the integration contour into the lower part of the $\lambda$-plane but then there is no contribution from Regge poles.

These arguments show that the Regge poles contribute to the nearside amplitude when $\theta < \theta_g$ and to the farside amplitude as illustrated in fig. 6.2. The nearside amplitude $f_d^-(\theta)$ in the shadow region $\theta > \theta_g$ has no connection with Regge poles. The Regge pole contribution in Frahn's theory can be calculated by substituting the formula (6.13) for $F(\chi)$ into (5.8) and (5.9). Here we prefer to obtain it directly from the integrals (4.17)

$$f_0^{\pm}(\theta) = \frac{1}{ik}(2\pi\sin\theta)^{-\frac{1}{2}} \int_0^\infty \lambda^{\frac{1}{2}} d\lambda S(\lambda) \exp i(\pm\lambda\theta \mp \tfrac{1}{4}\pi)$$

When $\theta_g > \theta$ the nearside amplitude has a Coulomb contribution from the Coulomb saddle point and one from the Regge pole

$$f_0^-(\theta) \approx f_C(\theta) - C_1(\sin\theta)^{-\frac{1}{2}} \exp i(2\sigma(\lambda_1) - \lambda_1\theta) \tag{6.14}$$

where

$$C_1 = (\beta_1/k)(2\pi\lambda_1)^{\frac{1}{2}} e^{\frac{1}{4}i\pi}$$

The farside amplitude has only a pole contribution

$$f_0^+(\theta) \approx iC_1(\sin\theta)^{-\frac{1}{2}} \exp i(2\sigma(\lambda_1) + \lambda_1\theta) \tag{6.15}$$

The magnitude of the Regge pole contribution to (6.14) and (6.15) can be calculated approximately by expanding the Coulomb phase

$$2\sigma(\lambda_1) \approx 2\sigma(\operatorname{Re}\lambda_1) + i\bar{\theta}_g \operatorname{Im}\lambda_1$$

where $\bar{\theta}_g$ is the Coulomb scattering angle corresponding to the angular momentum $\operatorname{Re}\lambda_1$. It is

$$|\beta_1|(2\pi|\lambda_1|/\sin\theta)^{\frac{1}{2}} \exp\{-\operatorname{Im}\lambda_1(\bar{\theta}_g \mp \theta)\} \tag{6.16}$$

The angle dependence is determined by $\operatorname{Im}\lambda_1$ where $\lambda_1$ is the Regge pole which is nearest to the real $\lambda$-axis.

The back-angle cross-section can be calculated from the Regge pole contribution to the $n = 1$ term in (5.45). The resulting amplitude can be obtained from (5.49) by the replacement $\Delta \to -\beta_1$. The ratio of the cross-section at $\theta = \pi$ to the Rutherford cross-section is given by a formula similar to (5.51)

$$\sigma(\pi)/\sigma_R(\pi) \approx \left|\frac{4\pi\beta_1\lambda_1}{n}\right|^2 \exp\{-2\operatorname{Im}\lambda_1(\bar{\theta}_g + \pi)\} \tag{6.17}$$

Several authors (see Fuller & Moffa, 1977, Anni *et al.*, 1980) have discussed the Regge pole contribution to the scattering amplitude in a way which avoids the use of the Poisson formula and the Frahn approximations. Fuller and Moffa (1977) use the formula

$$f(\theta) \approx f_C(\theta) + A_\alpha P_\alpha(\cos\theta); \quad \theta < \theta_g \tag{6.18}$$

The Regge pole term proportional to $P_\alpha$ with $\alpha = \lambda_1 - \frac{1}{2}$ contains both the nearside and farside amplitudes. If $\theta$ is not too small the nearside part dominates, the Legendre function has the asymptotic form

$$P_\alpha(\cos\theta) \sim (2\pi/\lambda_1 \sin\theta)^{\frac{1}{2}} \exp i(-\lambda_1\theta + \tfrac{1}{4}\pi)$$

and (6.18) is equivalent to (6.14).

Lowdowne (1979) has given the following formula for the back angle cross-section

$$\frac{\sigma(\pi)}{\sigma_R(\pi)} = \left|\frac{2\pi\beta_1\lambda_1}{n}\right|^2 \frac{\exp(-2\operatorname{Im}\alpha\bar{\theta}_g)}{\sin^2(\pi\operatorname{Re}\alpha) + \sinh^2(\pi\operatorname{Im}\alpha)} \tag{6.19}$$

It is derived without using the Poisson sum formula and effectively sums all the terms in the Poisson series. In most applications $\pi\operatorname{Im}\alpha = \pi\operatorname{Im}\lambda_1 \gg 1$ and (6.19) is equivalent to (6.17).

Fuller & Moffa (1977) have shown how to use the simple exponential dependence of the pole contribution in (6.18) to extract the location $\lambda_1$

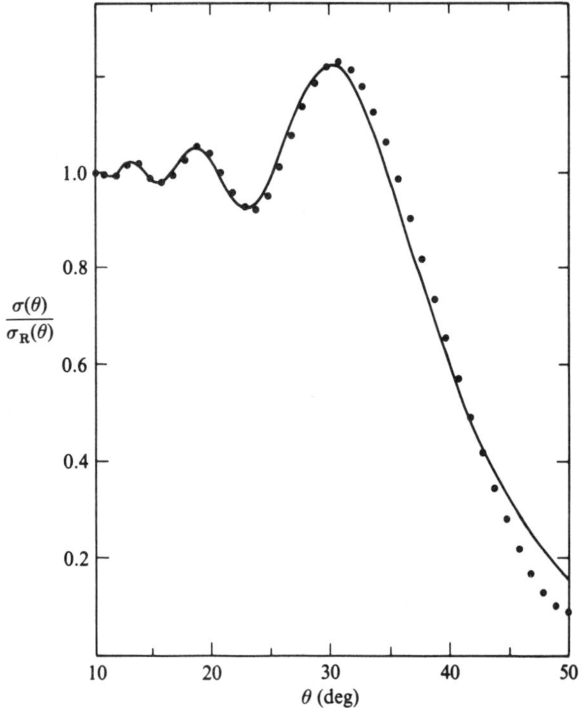


Fig. 6.4. Comparison of an $^{16}O + ^{48}Ca$ optical model calculation of $\sigma/\sigma_R$ with a Regge pole fit using equation (6.18). The solid curve is from the optical model and the dots are from the Regge pole fit. The energy is $E_{lab} = 56$ MeV and the Regge pole parameters are $A_1 = -2.33\exp(1.22i)$, $\lambda_1 = 30 + 4.12i$.

as well as the residue of the Regge pole. They take $A_1$ and $\lambda_1 = \alpha + \frac{1}{2}$ as parameters and choose them so that $|f(\theta)|^2$ is a good fit to the experimental cross-section. Fig. 6.4 shows an example for $^{16}O + ^{48}Ca$ scattering at $E_{lab} = 56$ MeV. The solid curve is the result of an optical model calculation of $\sigma/\sigma_R$ while the dots show the same quantity calculated from the Regge pole formula (6.18) with $A_\alpha = -2.33\exp(1.22i)$ and $\lambda_1 = 30 + 4.12i$. The Regge pole cross-section is almost the same as the optical model result up to the Blair quarter-point angle.

The E18 optical potential (Cramer *et al.*, 1976) has been used to describe $^{16}O + ^{28}Si$ scattering. At $E_{lab} = 55$ MeV it is an example of a potential which has an isolated Regge pole near the real $\lambda$-axis. Takemasa & Tamura (1978) made a numerical calculation of the location of this Regge pole and the corresponding residue. They found $\lambda_1 = 24.29 + 4.20i$. Anni *et al.* (1980) used the results of Takemasa & Tamura in the Regge pole formula

(6.18) and compared the result with the exact optical model amplitude. They found that the two amplitudes were accurately equal for $\theta < \theta_g$.

## 6.3 Orbiting and barrier top resonances

Consider an effective potential with a barrier and assume that there is no absorption at the position of the barrier. The height of the barrier $V_B(\lambda)$ depends on the angular momentum $\lambda$ and there is a critical angular momentum $\lambda_0$ for which the barrier height is equal to the energy of relative motion

$$V_B(\lambda_0) = E \tag{6.20}$$

If $\lambda < \lambda_0$ there is enough energy to get over the barrier into the internal region. When $\lambda > \lambda_0$ there is reflection at the barrier. The classical deflection function for such a potential is shown in fig. 6.5. The singularity at $\lambda = \lambda_0$ corresponds to classical orbiting. When $\lambda - \lambda_0$ is small the relative coordinate $r$ remains near the barrier position $r_B$ for a long time and the deflection angle is large. The deflection function has two branches: The one for $\lambda < \lambda_0$ corresponds to a trajectory which passes over the barrier, the other for $\lambda > \lambda_0$ to a trajectory reflected at the barrier. For $|\lambda - \lambda_0|$ small the classical scattering angle is

$$\Theta(\lambda) \approx \theta_1 + \theta_0 \ln(\lambda - \lambda_0); \qquad \lambda > \lambda_0 \tag{6.21}$$

$$\Theta(\lambda) \approx \theta_2 + 2\theta_0 \ln(\lambda_0 - \lambda_1); \quad \lambda < \lambda_0 \tag{6.22}$$

When $\Theta(\lambda) > 0$ then the observation angle $\theta = \Theta(\lambda)$ and when $\Theta(\lambda) < 0$ the relation between $\theta$ and $\Theta(\lambda)$ is given by equation (3.54) and illustrated in fig. 6.5. Due to the orbiting many deflection angles correspond to the same angle of observation.

In this section we consider the case where there is no absorption near and outside the barrier at $r_B$ but where there is a strong absorption inside the barrier. The effect is to eliminate the internal branch ($\lambda < \lambda_0$) of the deflection function in fig. 6.5, while the external branch ($\lambda > \lambda_0$) is unaffected. Ford & Wheeler (1959) have given a formula for the classical orbiting cross-section which can be obtained from the general classical cross-section formula (4.1) and the deflection function (6.21) which can also be written as

$$\lambda - \lambda_0 = \exp\{(\Theta(\lambda) - \theta_1)/\theta_0\}$$

The cross-section formula has several branches because the relation (3.54) between deflection angle and observation angle is not one-to-one. When $\theta_1 > \theta > 0$ the cross-section is nearside

$$\sigma^-(\theta) = (\lambda_0/k^2\theta_0 \sin\theta)\exp\{(\theta - \theta_1)/\theta_0\} \tag{6.23}$$

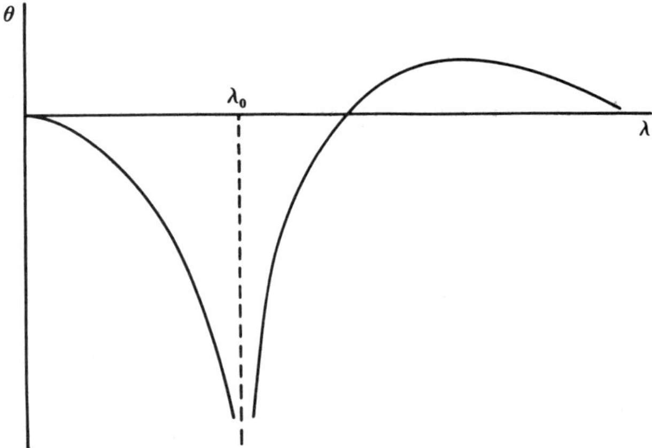


Fig. 6.5. The classical deflection function showing the orbiting effect; $\lambda_0$ is the orbiting angular momentum.

When $0 > \Theta(\lambda) > -\pi$ we have $\theta = -\Theta(\lambda)$ and a farside cross-section

$$\sigma^+(\theta) = (\lambda_0/k^2\theta_0 \sin\theta)\exp\{-(\theta_1 + \theta)/\theta_0\} \tag{6.24}$$

When $-\pi > \theta(\lambda)$ there are further contributions which correspond to $m \neq 0$ (3.54). They are usually very small compared with (6.23) and (6.24) and need not be considered. The cross-sections $\sigma^-(\theta)$ and $\sigma^+(\theta)$ have an exponential dependence on angle with a decay constant determined by $\theta_0$ (called the life angle by Fuller, 1973). This decay angle depends on the curvature of the top of the barrier and is larger for a thicker barrier. It is given by an approximate formula

$$\theta_0 \approx \omega_0/\omega_{\mathrm{B}} \tag{6.25}$$

where $\omega_0 = \hbar\lambda_0/\mu r_{\mathrm{B}}^2$ is the orbiting angular velocity and

$$\omega_{\mathrm{B}} = |V_{\mathrm{B}}''/\mu|^{\frac{1}{2}}$$

is the angular frequency of simple harmonic motion in a potential obtained by turning the barrier upside down. The quantity $\omega_{\mathrm{B}}^{-1}$ is the mean time to stay on top of the barrier and $\theta_0$ is the angle turned through in that time.

Friedman & Goebel (1977) have shown that there is a very close connection between classical orbiting and quantum mechanical barrier penetration. We now discuss this point. The discussion starts with equation (6.1) which gives the effects of barrier penetration on the nuclear partial wave amplitude $S_{\mathrm{n}}(\lambda)$. If the absorption for $r < r_{\mathrm{B}}$ is large the action $W_{32}$ has a large positive imaginary part and the exponentials $\exp(2iW_{32})$ in (6.1) can be neglected so that

$$S_n(\lambda) \approx S_B(\lambda) = \exp(2i\delta_1(\lambda))/N(i\beta) \tag{6.26}$$

From equation (E.12) in the appendix

$$1/N(i\beta) = (2\pi)^{-\frac{1}{2}}\Gamma(\tfrac{1}{2} + i\beta)\exp\{\tfrac{1}{2}\pi\beta - i(\beta \ln \beta - \beta)\}$$

and the integral (6.3) for $\beta$ can be evaluated approximately to give

$$\beta \approx (V_B(\lambda) - E)/\hbar\omega_B$$

Using equation (6.20) for the orbiting condition $\beta$ can also be written in terms of $\lambda$

$$\beta \approx (\lambda - \lambda_0)\theta_0 \tag{6.27}$$

where $\theta_0$ is given in (6.25). When $\beta$ is large $N \simeq 1$ and (6.26) simplifies to the WKB result

$$S_n(\lambda) \approx \exp(2i\delta_1(\lambda)) \tag{6.28}$$

where $\delta_1(\lambda)$ is the semi-classical phase calculated for the outer turning point $r_1$ in fig. 6.1, but near $\lambda \approx \lambda_0$ there is an important difference. The two sides of equation (6.28) have a different analytic structure as a function of the complex variable $\lambda$. Near $\lambda = \lambda_0$ the WKB phase $\delta_1(\lambda)$ is

$$2\delta_1(\lambda) = g(\lambda) + (\beta \ln \beta - \beta)$$

where $g(\lambda)$ is regular at $\lambda = \lambda_0$ and the term $\beta \ln \beta$ has a branch point there. Using (6.27) for $\beta$ the classical deflection function is

$$\Theta(\lambda) = 2\delta_1'(\lambda) = g'(\lambda) + \theta_0 \ln\{(\lambda - \lambda_0)\theta_0\}$$

Comparing with (6.21) we see that the logarithmic singularity in $\Theta(\lambda)$ is related to the branch point in $\delta_1(\lambda)$.

In equation (6.26) the logarithmic singularity associated with $2\delta_1(\lambda)$ is cancelled by a corresponding singularity in $N$ and the branch cut associated with $\lambda_0$ is replaced by a series of poles which are the poles of the gamma function in $N$. The pole positions are given by

$$\left.\begin{aligned} &(\tfrac{1}{2} + i\beta) = -n \\ \text{or}\quad &\lambda = \lambda_0 + i/2\theta_0, \quad \lambda_0 + 3i/2\theta_0, \cdots \end{aligned}\right\} \tag{6.29}$$

These poles are Regge poles associated with 'barrier top resonances' (Friedman & Goebel, 1977) and represent the quantal analogue of the classical orbiting phenomenon. If we calculate the contribution of the nearest Regge pole to the cross-section by the methods of section 6.2 and use the result of equation (6.16) we find that the Regge pole cross-section has the same exponential angle dependence as the classical orbiting cross-section (6.23) and (6.24) with the same decay angle $\theta_0$.

If the optical potential has a small imaginary part $W_B$ at the barrier there are two reasons for damping of the orbiting motion or, equivalently, the barrier top resonances. The projectile can fall off the barrier or it can be absorbed. The total width for absorption is a sum of contributions from each of these processes

$$\Gamma = \hbar\omega_B + 2W_B$$

($2W_B/\hbar$ is the probability per unit time for absorption as shown in section 2.2). Equation (6.22) for the decay angle $\theta_0$ is modified to

$$1/\theta_0 = \omega_B/\omega_0 + 2W_B/\hbar\omega_0 \tag{6.30}$$

The first barrier-top Regge pole is still determined by (6.29) with equation (6.30) for $\theta_0$. Equation (6.30) can also be derived by evaluating the integral (6.3) for $\beta$ in the presence of a weak imaginary potential.

The arguments in this section lead to the following qualitative conclusions. Barrier-top resonances are the quantum analogue of classical orbiting. Both give the same exponential dependence (6.23), (6.24) for the cross-section on angle. When the optical potential is real at the barrier the decay angle $\theta_0$ is given by equation (6.25). The decay angle is larger for a thicker barrier. When the optical potential has an imaginary part at the barrier the cross-section still has an exponential dependence with a smaller decay angle. The absorption dominates over orbiting when $W_B > \frac{1}{2}\hbar\omega_B$. A surface transparent potential has

$$W_B < \tfrac{1}{2}\hbar\omega_B \tag{6.31}$$

The farside amplitude $f_a^+(\theta)$ has an exponential angle dependence (6.16) for such a potential with decay constant determined by $\operatorname{Im}\lambda_1 = 1/(2\theta_0)$.

## 6.4 Barrier waves and internal waves

The deflection function for the potential discussed in section 6.3 has an external branch due to orbits with $\lambda > \lambda_0$ which are reflected at the barrier and an internal branch due to trajectories with $\lambda < \lambda_0$ which pass over the barrier and penetrate into the internal region. In section 6.3 we assumed that the absorption was strong inside the barrier and the effects of the internal branch were eliminated. If the absorption of the internal branch is not complete the trajectory can pass out over the barrier and contribute to the cross-section. A calculation similar to the one in section 6.3 gives a farside classical cross-section due to orbits with deflection angle in the range $0 > \Theta(\lambda) > -\pi$ which is approximately

$$\sigma^+(\theta) = \sigma_B^+(\theta) + \sigma_I^+(\theta)$$
$$= [\lambda_0/(k^2\theta_0 \sin\theta)](\exp\{-(\theta+\theta_1)/\theta_0\} + \tfrac{1}{2}C_I \exp\{-(\theta+\theta_2)/2\theta_0\}) \tag{6.32}$$

The first term in (6.32) is the same as (6.24) and is due to trajectories reflected at the barrier. The second term is due to internal trajectories. It is calculated from the deflection function (6.22). The coefficient $C_I$ measures the absorption of the internal trajectory. If $C_I = 0$ it is completely absorbed and (6.32) reduces to (6.21). The case $C_I = 1$ corresponds to no absorption. Notice that the internal contribution to the cross-section decreases more slowly with angle than the barrier contribution. This is because the internal trajectory encounters the barrier twice, once on the way in and again on the way out. It can give a significant part of the cross-section for large $\theta$ even if the absorption is quite strong.

The semi-classical theory gives a decomposition analogous to the classical result (6.32), but it is written in terms of amplitudes

$$f(\theta) = f_B(\theta) + f_I(\theta) \tag{6.33}$$

Here $f_B(\theta)$ is the amplitude of a wave reflected at the barrier and $f_I(\theta)$ is an internal amplitude. The cross-section corresponding to (6.33) is

$$\sigma(\theta) = \sigma_B(\theta) + \sigma_I(\theta) + 2\,\mathrm{Re}(f_B^*(\theta) f_I(\theta)) \tag{6.34}$$

This resembles the classical result (6.32). The new thing is the interference term between the barrier and internal amplitudes. The decomposition (6.33) was suggested by Brink, Grabowski & Vogt (1978) and the semi-classical derivation was made by Brink & Takigawa (1977).

The derivation of (6.33) starts with the semi-classical formula (6.1) for the nuclear potential wave amplitude. Equation (6.1) can be rewritten as

$$S_n(\lambda) = S_B(\lambda) + S_I(\lambda) \tag{6.35}$$

where $S_B(\lambda)$ is the barrier amplitude (6.26) which was discussed in section 6.3,

$$S_I(\lambda) = \frac{\exp(2i\delta_3(\lambda))}{N(N + \exp(2iW_{32}))} \tag{6.36}$$

is an internal amplitude and

$$\delta_3(\lambda) = \delta_1(\lambda) + i\pi\beta + W_{32} \tag{6.37}$$

is the WKB nuclear phase shift calculated with respect to the internal turning point in fig. 6.1. Expanding the denominator in (6.36) as a power

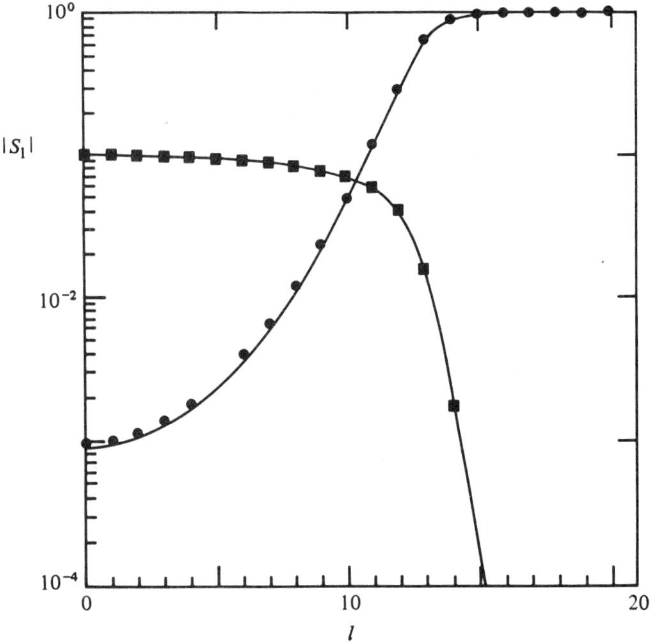


Fig. 6.6. The magnitudes of the barrier amplitude $|S_B|$ (dots) and the internal amplitude $|S_1|$ (squares) as a function of $l = \lambda - \frac{1}{2}$ for alpha particle scattering by $^{40}Ca$ at $E_{lab} = 20$ MeV. The dots and squares are calcuated from the semi-classical formulae (6.26) and (6.36) while the curves are calculated by the method of Albinski & Michel (section 6.5). The potential parameters are the same as for figs. 6.7 and 6.8 (from Albinski & Miche, 1982).

series in $N^{-1}\exp(2iW_{32})$ gives a multi-reflection series for $S_I(\lambda)$

$$S_I(\lambda) = \exp(2i\delta_3)/N^2(i\beta) + \cdots \tag{6.38}$$

The first term in the expansion for the internal amplitude is the quantal analogue of the internal deflection function in fig. 6.5. Higher terms correspond to waves which suffer a number of internal reflections before they emerge. In nuclear applications they are strongly damped because of absorption and give a negligible contribution to the cross-section. The effects of the leading term in (6.38) are important for $\alpha$-particle elastic scattering and also for some other light systems such as $^{16}O + {}^{28}Si$. The internal amplitude $S_I$ gives no observable contributions for heavier systems.

Fig. 6.6 is a sketch of the magnitude of $S_B$ and $S_I$ as a function of $l$. The barrier amplitude tends to zero for $\lambda < \lambda_0$ and the internal amplitude tends to zero for $\lambda > \lambda_0$. This is consistent with the association of $S_B$ with the external branch of the classical deflection function in fig. 6.5 and $S_I$ with the internal branch. Classically there is a sudden transition at $\lambda_0$. In the quantal theory there is a smoothing due to barrier penetration effects and

$$|S_B|^2 \approx [1 + \exp\{2\pi\theta_0(\lambda_0 - \lambda)\}]^{-1} \tag{6.39}$$

The width of the transition region is determined by $\theta_0$.

We conclude this section with a discussion of the semi-classical Regge poles of the amplitude (6.1). Anni & Renna (1981*a*, *b*) have discussed the singularities of $S_n(\lambda)$ as a function of the complex variable $\lambda$. The action integrals $\delta_1(\lambda)$ and $W_{32}(\lambda)$ in (6.1) have logarithmic singularities at the orbiting angular momentum $\lambda_0$. These cancel with corresponding singularities in the barrier amplitude (6.24) as discussed in section 6.3. They also cancel on the right-hand side of (6.36) and in every term of the multiple reflection series (6.38). Thus equations (6.1), (6.36) and each term in (6.38) are analytic functions at $\lambda = \lambda_0$.

Semi-classical Regge poles are poles of $S_n(\lambda)$ or zeros of the denominator in (6.1)

$$N(i\beta) + \exp(2iW_{32}) = 0 \tag{6.40}$$

In contrast the terms of the multiple reflection series have poles corresponding to the poles of the gamma function in formula (6.26a) for $N^{-1}(i\beta)$. These are the barrier top resonances (6.29). It is noteworthy that, although equation (6.35) with $S_I$ given by the first term in (6.38) can give a good approximation to (6.1) for real $\lambda$, it has a completely different singularity structure in the complex $\lambda$-plane. To avoid confusion between the poles of $S_n(\lambda)$ and the poles of the terms in the multi-reflection series (6.38) di Salvo & Viano (1977) have proposed that the name 'Regge poles' should be reserved for the poles of $S_n(\lambda)$. The poles of $S_B$ and of the terms in the multi-reflection series they call 'Sommerfeld poles'. We use their terminology in this book. Debye (1908) used a multi-reflection series like (6.38) in a study of scattering of light from a refracting cylinder. Nussenzweig (1979) calls this a 'Debye expansion' and the poles of the terms 'Regge–Debye' poles.

Anni & Renna (1981*a*, *b*) have shown that the imaginary part of the optical potential provides a link between the Regge and Sommerfeld poles. If the strength of the imaginary part of the potential is increased from zero the Regge poles closest to the real $\lambda$-axis move upwards. One of them approaches the location of the first Sommerfeld pole at $\beta = \frac{1}{2}i$. When $W_0$ is increased further this pole does not follow the other Regge poles in their upward motion and begins to behave like a Sommerfeld pole. In this limit the $S$-matrix is well approximated by the barrier $S$-matrix $S_B(\lambda)$ as discussed in section 6.3.

Takemasa & Tamura (1978) have calculated the positions of Regge poles for a number of potentials E18, GK and SD used to fit $^{16}O + {}^{28}Si$ experimental data, and Anni & Renna have compared the Regge pole positions with those of the Sommerfeld poles. They find that for the E18 potential the nearest Regge pole almost coincides with the first Sommerfeld pole. In the case of the GK and SD potentials, which are less strongly absorbing, there is no correspondence between any of the Regge poles and the Sommerfeld pole. Anni & Renna (1981*b*) and Anni *et al.* (1978) also show that for the potentials SK and GR the back angle cross-section is due in the Regge description to the contribution of one or two Regge poles, and to the nearest Sommerfeld pole in the multi-reflection description. These potentials provide a clear example of how different representations of the same phenomenon are possible in the transition region between the quantal and classical regimes.

## 6.5 Anomalous large angle scattering

Elastic $\alpha$-particle scattering from calcium isotopes has been thoroughly investigated because of the large backward enhancement seen in some experimental angular distributions (see Delbar *et al.*, 1978). This 'anomalous large angle scattering' (ALAS) was a puzzle for many years. There is a marked variation in the cross-section from one nucleus to another. Angular distributions have a very complicated structure and change rapidly with energy. We know now that ALAS is a manifestation of internal-barrier wave interference and that the target dependence is due to a marked variation of the absorption from one nucleus to another. ALAS has also been observed in other light systems, for example $\alpha + {}^{16}O$ (Michel *et al.*, 1984) and $^{6}Li + {}^{16}O$ (Bassani *et al.*, 1972). The case of $^{16}O + {}^{28}Si$ will be discussed in section 6.6. Michel & Vanderpoorten (1977) made the first successful optical model analysis of a $\alpha + {}^{40}Ca$ scattering over a range of energies. Careful systematic experiments of Delbar *et al.* (1978) revealed a regular pattern in the ALAS phenomenon.

Fig. 6.7 shows a typical ALAS cross-section. It is the angular distribution for scattering of $\alpha$-particles by $^{40}Ca$ at $E_{lab} = 29$ MeV. The ratio $\sigma(\theta)/\sigma_R(\theta) > 1$ for backward angles. The curve in fig. 6.7 is calculated with an optical potential found by Delbar *et al.* and gives a good fit to the experimental data. The form factor of both the real and imaginary parts is a square of a Woods–Saxon (equation (2.12) with $\nu = 2$) and the para-

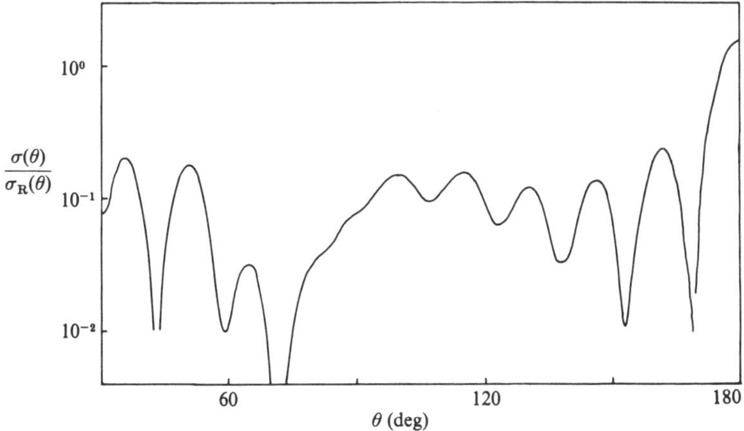


Fig. 6.7. Angular distribution for scattering of alpha particles by $^{40}Ca$ at $E_{lab} =$ 29 MeV. The potential parameters are given in the text.

meters are $V_0 = 188.92$ MeV, $W_0 = 11.342$ MeV, $R = 4.685$ fm, $R_w = 6$ fm, $a_v = 0.645$ fm, $a_w = 0.5$ fm.

The optical glory effect (Nussenzweig, 1979) is a very complicated process because the amplitude has contributions from waves which suffer many internal reflections before emerging from the scatterer. Heavy-ion scattering is simpler because the absorption is relatively strong and components associated with multiple reflections are damped out. There are, however, contributions from one or several internal waves in systems showing ALAS. Brink & Takigawa (1977) separate the barrier and internal partial wave amplitudes by using the semi-classical formulae (6.26) and (6.38). These are substituted into separate partial series to give $f_B(\theta)$ and $f_I(\theta)$. The elastic cross-section obtained by this method compares well with the exact optical model cross-section. This proves that the semi-classical method is accurate. The barrier and internal cross-sections $\sigma_B(\theta)$ and $\sigma_I(\theta)$ each have a rather simple structure and the complicated form of the total cross-section (fig. 6.7) is due to interference between the barrier and internal amplitudes (equation 6.34).

The evaluation of the semi-classical amplitudes $S_B$ and $S_I$ is not difficult in principle. However, the method requires the location of the complex turning points $r_1$, $r_2$ and $r_3$ and the evaluation of several action integrals in the complex plane making its practical use inconvenient. Also it is restricted to potentials which are analytical functions of $r$, and thus it does not allow the direct investigation of the interesting case of the folding model or of any potential which is supplied numerically. A simple approach would be to eliminate the internal wave by artificially enhancing the absorption inside the potential pocket, so that only the barrier wave survives. The internal amplitude $f_I(\theta)$ can then be calculated by subtracting the barrier part $f_B(\theta)$ from the full amplitude $f(\theta)$. Such an approach was used by Rowley *et al.* (1977) and Brink *et al.* (1978) but sometimes it does not work. Albinski & Michel (1982) developed a more flexible method which is based on the structure of the expressions (6.26) and (6.38) for $S_B$ and $S_I$. These authors consider a modification of the optical potential

$$V(r) \rightarrow V(r) + Kg(r) \tag{6.41}$$

where $K$ is a small complex number and $g(r)$ is an analytical complex function of $r$, which is negligibly small outside the interval between the interior turning points $r_3$ and $r_2$. Thus the modification $Kg(r)$ leaves the action integrals $\delta_1$ and $\beta$ unchanged while

$$W_{32} \rightarrow W_{32} + K\alpha$$

where $\alpha$ is a complex constant. Hence

$$S = S_B + S_I \rightarrow S_B + S_I e^{2K\alpha} \tag{6.42}$$

The method of Albinski & Michel is to perform three successive optical model calculations with the following potentials

$$V(r), \quad V(r) + Kg(r), \quad V(r) - Kg(r)$$

where $V(r)$ is the optical potential whose properties are to be investigated. According to (6.42) the corresponding amplitudes will be

$$S^0 = S_B + S_I$$
$$S^+ = S_B + S_I e^{2K\alpha}$$
$$S^- = S_B + S_I e^{-2K\alpha}$$

These are calculated numerically using an optical model programme, then $S_B$ and $S_I$ are found by solving the three equations to give

$$S_I = -\Delta^+\Delta^-/(\Delta^+ + \Delta^-), \quad S_B = S^0 - S_I$$

where $\Delta^{\pm} = S^{\pm} - S^0$. Albinski & Michel have done many tests and have compared the results with direct computation of the semi-classical formulae (6.26) and (6.38). They conclude that the method is reliable. Its appealing feature is the fact that it can be implemented by making a simple modification of a standard optical model code. In this sense it resembles Fuller's (1975) method for making the nearside–farside decomposition which was discussed in section 4.7.

Fig. 6.8 shows the barrier and internal cross-sections $\sigma_B(\theta)$ and $\sigma_I(\theta)$ cal-

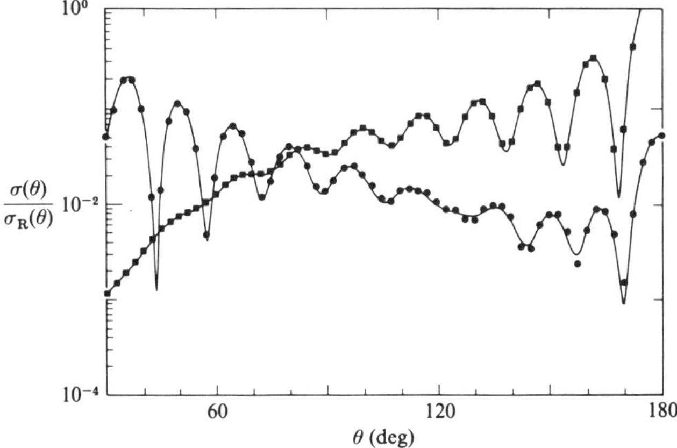


Fig. 6.8. The barrier cross-section $\sigma_B(\theta)$ (dots) and the internal cross-section $\sigma_I(\theta)$ (squares) for the scattering of $\alpha$-particles by $^{40}$Ca at $E_{lab} = 29$ MeV. The dots and squares are calculated from the semi-classical formulae (6.26) and (6.36) while the curves are calculated by the method of Albinski & Michel. The potential parameters are the same as for figs. 6.6 and 6.7 (from Albinski & Michel, 1982).

culated by Albinski & Michel for the $\alpha + {}^{40}$Ca cross-section shown in fig. 6.7. The internal cross-section $\sigma_I > \sigma_B$ if $> 80°$ and gives the dominant contribution at backward angles. Both $\sigma_B(\theta)$ and $\sigma_I(\theta)$ have a very regular angular dependence which can be understood by the concepts developed in chapters 4 and 5 and in section 6.3. The complex angular dependence in fig. 6.7 is due to interference of the amplitudes $f_B$ and $f_I$. Fig. 6.8 shows that the spacing between the oscillations in $\sigma_B(\theta)$ is somewhat smaller than for $\sigma_I(\theta)$. Extracting a characteristic angular momentum using equation (5.39) gives $\lambda_B = 13.8$ for the barrier cross-section and $\lambda_I = 11.8$ for the internal cross-section.

When the $\alpha$-particle energy is increased the barrier and internal components $\sigma_B(\theta)$ and $\sigma_I(\theta)$ evolve in a regular way. The characteristic angular momenta $\lambda_B$ and $\lambda_I$ increase as expected and the relative phase of $f_B$ and $f_I$ changes so that the gross structures in the total elastic cross-section which show clearly in fig. 6.7 move forward in angle. Classical orbiting is no longer possible for $\alpha$-particle energies greater than about 60 MeV and the amplitude $f(\theta)$ cannot be decomposed into barrier and internal components. At these higher energies the cross-sections begin to show the characteristic nuclear rainbow pattern which was discussed in section 4.8 and illustrated in fig. 4.8.

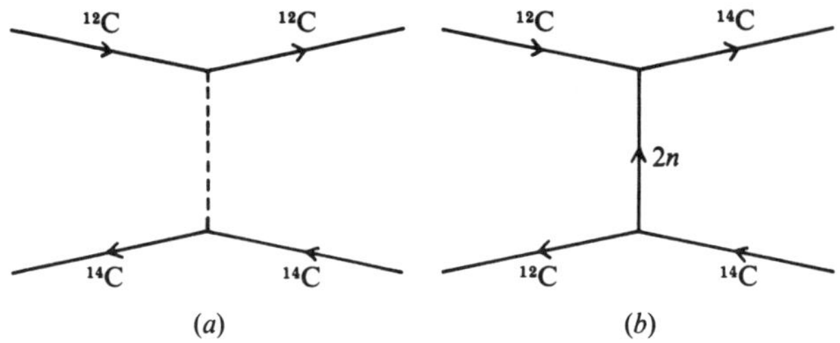


Fig 6.9. Reaction mechanisms in the elastic scattering of almost identical nuclei: (*a*) direct scattering, (*b*) elastic scattering.

## 6.6 Exchange scattering

Elastic scattering of pairs of similar nuclei like $^{12}C + ^{14}\mathrm{C}$ frequently do not have angular distributions which are characteristic of strong absorption models. We describe a model of 'elastic transfer' which can explain the observed effects. Fig. 6.9 illustrates the mechanism of elastic transfer for the $^{12}C + ^{14}\mathrm{C}$ system. Fig. 6.9(*a*) represents the normal direct component of elastic scattering and fig. 6.9(*b*) is the exchange component. In fig. 6.9(*b*) two neutrons are transferred out of the $^{14}\mathrm{C}$ and it becomes a $^{12}\mathrm{C}$ nucleus. The two neutrons are absorbed by the $^{12}\mathrm{C}$ and it is transformed into $^{14}\mathrm{C}$. If the resulting nuclei are in their ground states the final state is indistinguishable from elastic scattering. The forward-angle cross-section will be dominated by the direct component whereas the transfer mechanism will be stronger at backward angles. The scattering amplitude can be written as a sum of a direct part $f_\mathrm{D}$ and a transfer or exchange part $f_\mathrm{E}$

$$f(\theta) = f_\mathrm{D}(\theta) + f_\mathrm{E}(\pi - \theta) \tag{6.43}$$

The direct part has a partial wave expansion

$$f_\mathrm{D}(\theta) = \sum_l (2l+1) P_l(\cos\theta)(S_{\mathrm{D}l} - 1) \tag{6.44}$$

where $S_{\mathrm{D}l}$ is the direct partial wave amplitude. The exchange part has the partial wave expansion

$$\begin{aligned} f_\mathrm{E}(\pi - \theta) &= \sum_l (2l+1) P_l(\cos\theta)(-1)^l a_l \\ &= \sum_l (2l+1) P_l(\cos(\pi - \theta)) a_l \end{aligned} \tag{6.45}$$

where $a_l$ is an amplitude for two-neutron transfer in the partial wave $l$ and we have used $P_l(-x) = (-1)^l P_l(x)$. This amplitude is expected to be peaked near the grazing angular momentum and the enhancement at backward angles is a consequence of the factor $(-1)^l$ in equation (6.45).

The amplitude for any scattering showing elastic transfer effects will have the form (6.43) provided the nuclei involved have zero spin. In the case of non-zero spin the amplitudes will depend on spin quantum numbers as well as on the scattering angles.

Elastic transfer can be described by an optical potential which includes an exchange part

$$V = V_D(r) + V_E(r)P_M \tag{6.46}$$

where $P_M$ is the Majorana exchange operator. In a partial wave $l$ the exchange operator has the value $P_M = (-1)^l$ so the optical potential in states with even $l$ is $V_+(r)$ and in states with odd $l$ it is $V_-(r)$ where

$$V_{\pm}(r) = V_D(r) \pm V_E(r)$$

If the partial wave amplitudes corresponding to $V_{\pm}(r)$ are $S_{\pm l}$ then the total amplitude

$$\begin{aligned} f(\theta) = &\sum_{l\,\text{even}} (2l+1)P_l(\cos\theta)(S_{+l}-1) \\ &+ \sum_{l\,\text{odd}} (2l+1)P_l(\cos\theta)(S_{-l}-1) \end{aligned} \tag{6.47}$$

Direct calculation shows that equation (6.47) for $f(\theta)$ has the form (6.45) where

$$S_{Dl} = \tfrac{1}{2}(S_{+l} + S_{-l})$$
$$a_l = \tfrac{1}{2}(S_{+l} - S_{-l})$$

The direct and exchange components in (6.43) can also be written as

$$f_D(\theta) = \tfrac{1}{2}(f_+(\theta) + f_-(\theta))$$
$$f_E(\pi-\theta) = \tfrac{1}{2}(f_+(\pi-\theta) - f_-(\pi-\theta))$$

where $f_{\pm}(\theta)$ are the amplitudes calculated from the optical potentials $V_{\pm}(r)$.

A number of experiments have been made on elastic transfer (see Von Oertzen, 1970; Gubler *et al.*, 1977; Kobono *et al.*, 1980). Angular distributions show a characteristic structure due to interference of the direct and exchange amplitudes in equation (6.43). The semi-classical theory has been discussed by Frahn (1980*a*).

# 7

# Feynman's path integral in one dimension

## 7.1 Definition of the path integral

Feynman's original path integral approach to quantum mechanics was given in Lagrangian form. The propagator in coordinate space from an initial point $x_0$ to a final point $x_1$ after a time interval $T$ is expressed as a functional integral

$$K(x_1, x_0, T) = \int D[x(t)] \exp((i/\hbar) S[x(t)]) \tag{7.1}$$

where $S$ is the action functional

$$S[x(t)] = \int_0^T \mathscr{L}(x, \dot{x})\, dt = \int_0^T (\tfrac{1}{2} m \dot{x}^2 - V(x))\, dt \tag{7.2}$$

The path $x(t)$ satisfies boundary conditions

$$x(0) = x_0, \quad x(T) = x_1. \tag{7.3}$$

In (7.2) $\mathscr{L}$ is the classical Lagrangian. In this book $\mathscr{L}$ is always considered to have no explicit time dependence.

In the standard definition (Feynman & Hibbs, 1965) of the path integral the time interval $(0, T)$ is divided into $N$ equal subintervals $0 < t_1 < t_2 \ldots t_{N-1}$. Each path $x(t)$ is approximated by a polygon with vertices $x(t_i)$, $i = 1, \ldots, N-1$. The action functional becomes a function of the $N-1$ variables $x(t_i)$. The path integral is defined as a limit

$$K = \lim_{N \to \infty} (A_N)^N \int \exp((i/\hbar) S[x_1, x_2, \ldots, x_{N-1}]) dx_1, \ldots, dx_{N-1}$$

where the normalized factors

$$A_N = (2\pi i \hbar T / mN)^{\frac{1}{2}}$$

ensure the proper convergence of the limit as $N \to \infty$.

An alternative approach is to use the path expansion method introduced by Davison (1954) and developed by Levit & Smilansky (1977, *a*, *b*). Paths are expanded in a complete set of orthonormal functions. Let $\xi(t)$ be some particular path satisfying the boundary conditions (7.3) and let $u_\alpha(t)$ be a complete orthonormal set of functions defined in the interval $(0, T)$

$$\left.\begin{aligned}\int_0^T u_\alpha(t)u_\beta(t)\,dt &= \delta_{\alpha\beta}\\ u_\alpha(0) = u_\alpha(T) &= 0\end{aligned}\right\} \tag{7.4}$$

Then a general path $x(t)$ satisfying (7.3) can be expressed as

$$x(t) = \xi(t) + \sum_{\alpha=1}^{\infty} a_\alpha u_\alpha(t) \tag{7.5}$$

Define an approximation to $x(t)$ by truncating the series after $N$ terms

$$x_N(t) = \xi(t) + \sum_{\alpha=1}^{N} a_\alpha u_\alpha(t)$$

Then the action $S[x_N(t)]$ is a function of the $N$ parameters $a_1, a_2, \ldots, a_N$. Define the propagator as

$$K(x_1, x_0, T) = \lim_{N\to\infty} K_N(x_1, x_0, T)$$

where

$$K_N = \left[\frac{m}{2\pi i\hbar T}\right]^{\frac{1}{2}} \int \exp((i/\hbar)S[\xi(t), a_1, \ldots, a_N])\frac{da_1}{A_1}, \ldots, \frac{da_N}{A_N} \tag{7.6}$$

Levit and Smilansky (1977*a*, *b*) argue that the normalization factors $A_\alpha$ are independent of the choice of $\xi(t)$ and the $u_\alpha(t)$ and of the Lagrangian of the system. They choose the $A_\alpha$ by studying free particle motion. To find $A_\alpha$ take the $u_\alpha(t)$ to be eigenfunctions of the eigenvalue problem

$$m(d^2u_\alpha/dt^2) + \lambda_0^\alpha u_\alpha = 0 \tag{7.7}$$

with $u_\alpha(0) = u_\alpha(T) = 0$ so that

$$u_\alpha(t) \propto \sin(\pi\alpha t/T), \quad \lambda_0^\alpha = m(\pi\alpha/T)^2 \tag{7.8}$$

Levit (1978) shows that the result for $K$ is independent of the choice of the reference path $\xi(t)$ but the calculations are simplest when $\xi(t)$ is the classical path

$$\xi_0(t) = x_0 + (x_1 - x_0)t/T$$

Then

$$S[\xi_0, a_1, \ldots, a_N] = S_0[\xi_0(t)] + (1/2\hbar)\sum_{\alpha=1}^{N} \lambda_0^\alpha a_\alpha^2$$

with

$$S_0[\xi_0(t)] = m(x_1 - x_0)^2/2\hbar T$$

and

$$K_N(x_1, x_0, T) = \left[\frac{m}{2\pi i\hbar T}\right]^{\frac{1}{2}} \exp((i/\hbar)S_0[\xi_0(t)]) \times \prod_{\alpha=1}^{N} A_\alpha^{-1} \int_{-\infty}^{\infty} da\alpha \exp(i\lambda_0^\alpha a_\alpha^2/2\hbar) \tag{7.9}$$

The term outside the product is the free particle propagator so that (7.9) is the exact free particle propagator for every $N$ provided the $A_\alpha$ are chosen so that

$$A_\alpha = \int_{-\infty}^{\infty} da_\alpha \exp(i\lambda_0^\alpha a_\alpha^2/2\hbar) = [2\pi i\hbar/\lambda_0^\alpha]^{\frac{1}{2}} \tag{7.10}$$

We conclude this section by noting that the path integral approach can also be formulated in phase space. This form was introduced by Garrod (1966) and Davies (1963) and the path expansion method was applied to this formulation by Levit & Similansky (1977*b*). The propagator is written as a path integral over coordinates and momenta

$$K(x_1, x_0, T) = \int D[x(t)]D[p(t)]\exp((i/\hbar)S[x(t), p(t)])$$

$$S[x(t), p(t)] = \int_0^T ((p\dot{x} - H(p, x))\, dt$$

## 7.2 Properties of the action on classical paths

Let $S(x_1, x_0, T)$ be the action (7.2) calculated along a path satisfying the classical equations of motion

$$\frac{d}{dt}\left(\frac{\partial \mathscr{L}}{\partial \dot{x}}\right) - \frac{\partial \mathscr{L}}{\partial x} = 0 \tag{7.11}$$

with boundary conditions $x(0) = x_0, x(T) = x_1$. This action function satisfies three relations which are needed in subsequent sections

$$\partial S/\partial x_1 = p_1 \tag{7.12}$$

$$\partial S/\partial x_0 = -p_0 \tag{7.13}$$

$$\partial S/\partial t = -E \tag{7.14}$$

Here $p_0$ and $p_1$ are the momenta on the classical path at $t = 0$ and $t = T$ respectively, while $E$ is the total energy of the path. The first two relations can be obtained by making a general variation of the path. The change in $S$ is

$$\delta S = \int_0^T dt\delta x(t)\left(\frac{\partial \mathscr{L}}{\partial x} - \frac{d}{dt}\frac{\partial \mathscr{L}}{\partial \dot{x}}\right) + p_1\delta x_1 - p_0\delta x_0$$

Here $\delta x_0$ and $\delta x_1$ are the variations of the path end points. The first integral is zero if the path satisfies (7.11) and

$$\delta S = p_1\delta x_1 - p_0\delta x_0 \tag{7.15}$$

which yields the relations (7.12) and (7.13). The last relation can be obtained by making a simultaneous variation $\delta T$ of the time integral $T$ and $\delta x_1$ of $x_1$ in such a way that

$$\delta x_1 = \dot{x}_1 \delta T \tag{7.16}$$

where $\dot{x}_1$ is the velocity on the classical path when $t = T$. In other words, $x_1 + \delta x_1$ is obtained from $x_1$ by moving along the classical path in question. Then

$$\begin{aligned}\delta S = \mathscr{L}(x_1, \dot{x}_1)\delta T &= (\partial S/\partial x_1)\delta x_1 + (\partial S/\partial T)\delta T \\ &= (p_1 \dot{x}_1 + \partial S/\partial T)\delta T\end{aligned}$$

using (7.16). Hence

$$\partial S/\partial T = \mathscr{L}(x_1, \dot{x}_1) - p_1 \dot{x}_1 = -E$$

where $E$ is the energy at the end point of the orbit. But the energy is constant along the path and (7.14) is proved.

The second derivative of $S$ can be calculated by noting that there is no change in the energy $E(x_1, x_0, T)$

$$\delta E = 0 = (\partial E/\partial x_1)\dot{x}_1 \delta T + (\partial E/\partial T)\delta T$$

if $x_1$ and $T$ are changed so that (7.16) holds.
Hence

$$(\partial^2 S/\partial T^2) = -(\partial E/\partial T) = (\partial E/\partial x_1)\dot{x}_1$$

But

$$E = p_0^2/2m + V(x_0)$$

so that

$$(\partial E/\partial x_1) = (p_0/m)(\partial p_0/\partial x_1) = \dot{x}_0(\partial p_0/\partial x_1)$$

Combining these results gives

$$(\partial^2 S/\partial T^2) = \dot{x}_0 \dot{x}_1 (\partial p_0/\partial x_1) \tag{7.17}$$

This result is needed in section 7.4.

## 7.3 The primitive semi-classical approximation

The semi-classical limit of the propagator is obtained by evaluating the path integral (7.1) using asymptotic methods. The basic assumption behind these methods is that the most important contributions to the path integral come from the neighbourhood of classical paths

$$x(t) = x_c^\alpha(t), \quad \alpha = 1, 2 \ldots$$

for which the action integral is stationary

$$\delta S[x_c^\alpha(t)] = 0 \tag{7.18}$$

These paths satisfy classical equations of motion and the boundary conditions (7.3) at $t=0$ and $t=T$. In classical mechanics a solution is normally specified by giving initial conditions and a unique classical path results. In the present application the use of boundary conditions at $t=0$ and $t=T$ leads in many practical cases to more than one classical solution.

If only one classical path exists then we shall show that approximating the path integral leads to the result

$$K(x_1,x_0,T)=e^{-\frac{1}{2}i\pi\nu}\left|\frac{i}{2\pi\hbar}\frac{\partial^2 S}{\partial x_0\partial x_1}\right|^{\frac{1}{2}}\exp((i/\hbar)S(x_1,x_0,T)) \tag{7.19}$$

where $S(x_1,x_0,T)$ is the action calculated along the classical path. The integer $\nu$ is usually called a Morse index and will be discussed later. The relations (7.12) and (7.13) permit the pre-exponential factor to be written in a number of different ways, because

$$\frac{\partial^2 S}{\partial x_0\partial x_1}=-\frac{\partial p_0}{\partial x_1}=\frac{\partial p_1}{\partial x_0} \tag{7.20}$$

In problems in more than one dimension the generalization of this factor is frequently called a van Vleck (1928) determinant.

The result (7.19) can be obtained by the path expansion method. There is a natural choice for the reference path and for the set $u_\alpha(t)$ of orthogonal functions to be used in the method. Take the reference path to be the classical path $x_c(t)$ linking the boundary points $x_0$ and $x_1$ and write the general path as

$$x(t)=x_c(t)+\eta(t) \tag{7.21}$$

Then $S$ is expanded about the classical path to second order in $\eta(t)$

$$S[x_c(t)+\eta(t)]\simeq S[x_c(t)]+\delta^2 S_c \tag{7.22}$$

The first order variation is zero because of the stationary condition (7.18). The second order variation is

$$\delta^2 S_c=\int_0^T(\tfrac{1}{2}m\dot{\eta}^2+\tfrac{1}{2}R(t)\eta^2)dt \tag{7.23}$$

where $R(t)=-V''(x_c(t))$. The orthogonal functions $u_\alpha(t)$ of the path expansion method are chosen to be eigenfunctions of

$$-m(d^2u_\alpha/dt^2)+R(t)u_\alpha=\lambda^\alpha u_\alpha \tag{7.24}$$

with boundary conditions (7.4). If $\eta(t)$ is expanded in terms of the $u_\alpha$ as

$$\eta(t)=\sum_\alpha a_\alpha u_\alpha(t)$$

then to second order in the $a_\alpha$ the action is

$$S[x(t)] = S[x_c(t)] + \tfrac{1}{2}\sum_\alpha \lambda^\alpha a_\alpha^2 \tag{7.25}$$

If this expression is used in the formula (7.6) for the propagator and the gaussian integrals are calculated one obtains the semi-classical expression for the propagator

$$K(x_1, x_0, T) = \left[\frac{m}{2\pi i\hbar T}\right]^{\frac{1}{2}} \exp((i/\hbar)S(x_1, x_0, T))\prod_{\alpha=1}^{\infty}(\lambda_0^\alpha/\lambda^\alpha)^{\frac{1}{2}} \tag{7.26}$$

In (7.26) $\lambda_0^\alpha$ are the eigenvalues of the free particle problem (7.7) which corresponds to putting $R(t) = 0$ in (7.24).

All the eigenvalues $\lambda_0^\alpha$ of the free particle motion problem (7.7) are positive. Some of the eigenvalues of (7.24) may be negative. The Morse index $\nu$ in (7.19) is equal to the number of negative eigenvalues of (7.24).

Levit & Smilansky have shown how to calculate the product on the right-hand side of (7.26). They consider the equation

$$-m(d^2\eta/dt^2) + \mu R(t)\eta = 0 \tag{7.27}$$

where $\mu$ is a number between zero and unity and look for solutions $\eta(\mu, t)$ which satisfy

$$\eta(\mu, t) = 0, (d\eta(\mu, t)/dt) = 1 \quad \text{when } t = 0 \tag{7.28}$$

They show that

$$\prod_{\alpha=1}^{\infty}(\lambda^\alpha(\mu)/\lambda^\alpha(0)) = \eta(\mu, T)/\eta(0, T) \tag{7.29}$$

The free particle case corresponds to $\mu = 0$ and our case to $\mu = 1$. Hence the product in (7.26) is

$$\prod_{\alpha=1}^{\infty}(\lambda_0^\alpha/\lambda^\alpha) = \eta(0, T)/\eta(1, T)$$

Because of the initial condition (7.28)

$$\eta(1, T) = m(\partial x_1/\partial p_0), \quad \eta(0, T) = T$$

and combining these results in (7.26) gives (7.19). Equation (7.19) was derived assuming that there is only one classical path satisfying the boundary conditions (7.3). If several classical paths satisfying the boundary conditions exist and if they are well separated, each path gives contributions like (7.19) and the total propagator is a sum of contributions $K$, one from each classical path

$$K(x_1, x_0, T) = \sum_\alpha K^\alpha(x_1, x_0, T) \tag{7.30}$$

If two of the paths are close together then the approximation (7.22) of expanding the action up to second order about each path may not be accurate enough. In such a case it may be possible to use a uniform approximation to give a good result (see appendix C). Formula (7.30) shows that the contributions from different classical paths interfere and it is necessary to have the relative phases of the $K^\alpha$ correctly. This is the reason why the Morse index is important.

## 7.4 The energy representation

For many applications we need the propagator $G(E)$ in the energy representation. It was defined as the Fourier transform of $K(x_1, x_0, T)$

$$G(x_1, x_0, E) = 1/(i\hbar) \int_0^\infty dT K(x_1, x_0, T) \exp(iET/\hbar) \tag{7.31}$$

Derivations of the semi-classical formula for the propagator in the energy representation have been given by Gutzwiller (1967), Pechukas (1969) and other authors. The procedure usually followed consists of two steps. First, one obtains a semi-classical expression for the time-dependent propagator and afterwards evaluates the integral over $T$ in (7.31) by the saddle point approximation (SPA). This procedure gives the correct result for the magnitude of $G(E)$; but it may not always be possible to get the phase correctly. Möhring *et al.* (1980) show how to obtain the correct phase by calculating the path integral for $K$ and the time integral for $G$ in one step. Here we follow the more conventional approach while recognizing its limitations.

Using the semi-classical formula for $K$ from (7.19) and the definition of $G$ in (7.31) we get

$$G(x_1, x_0, E) = \frac{1}{i\hbar}\left[\frac{1}{2\pi i\hbar}\right]^{\frac{1}{2}} \int_0^\infty dT \left(\frac{\partial p_0}{\partial x_1}\right)^{\frac{1}{2}} \times \exp((i/\hbar)(S(x_1, x_0, T) + ET)) \tag{7.32}$$

In evaluating the integral by SPA the saddle point $T_s$ is given by

$$\frac{\partial S}{\partial T} = -E \tag{7.33}$$

The pre-exponential factor is assumed to be slowly varying and is evaluated at the saddle point $T_s$. The derivative $-(\partial S/\partial T)$ is the energy of the classical path with boundary conditions $x(0) = x_0, x(T) = x_1$ (see section 7.2). Thus (7.32) has the following meaning: Take the classical path with energy $E$

which starts at $x_0$ and ends at $x_1$. The time taken to move from $x_0$ to $x_1$ is $T_s$. The saddle point evaluation of the integral (see appendix B) gives

$$G(x_1, x_0, E) = \frac{1}{i\hbar}\left[\frac{1}{\hbar}\left(\frac{\partial p_0}{\partial x_1}\right)\bigg/\frac{\partial^2 S}{\partial T^2}\right]^{\frac{1}{2}} \cdot \exp((i/\hbar)W(x_1, x_0, E) - i\chi) \tag{7.34}$$

The phase $\chi$ depends on the sign of the second derivative of $S$ at the saddle point

$$\begin{aligned} \chi &= 0 \quad \text{if}\,(\partial^2 S/\partial T^2)_s > 0 \\ \chi &= \tfrac{1}{2}\pi \quad \text{if}\,(\partial^2 S/\partial T^2)_s < 0 \end{aligned} \tag{7.35}$$

In (7.34)

$$\begin{aligned} W(x_1, x_0, E) &= S(x_1, x_0, T_s) + ET_s \\ &= \int_0^{T_s} (p\dot{x} - H)dt + ET_s \\ &= \int_0^{T_s} p\dot{x}dt = \int_{x_0}^{x_1} pdx \end{aligned} \tag{7.36}$$

Note that the dependence on $T$ has disappeared from formula (7.36) for $W$. It remains to simplify the pre-exponential factor. This can be done by using (7.17) and the result is

$$G(x_1, x_0, E) = \frac{1}{i\hbar}(\dot{x}_0\dot{x}_1)^{-\frac{1}{2}} \exp\left((i/\hbar)\int_{x_0}^{x_1} pdx - i\chi\right) \tag{7.37}$$

If there are several saddle points in the time integration the total $G$ would be a sum of contributions like (7.37).

## 7.5 The momentum representation

Usually the incident momentum is specified in a scattering problem. For this reason it is also useful to study propagators like $K(x_1, p_0, T)$ which correspond to propagation from an initial momentum $p_0$ at $t = 0$ to a final position $x_1$ at $t = T$. The potential $V(x) \to 0$ as $x \to \pm\infty$ in a scattering problem. If $T$ is large enough so that the projectile is outside the range of the potential at $t = 0$ $p_0^2/2m = E$. Hence, there is a connection between the energy representation and the momentum representation when $T$ is large.

The propagator $K(x_1, p_0, T)$ is defined by

$$K(x_1, p_0, T) = \int_{-\infty}^{\infty} dx_0 \exp(ip_0 x_0/\hbar) K(x_1, x_0, T) \tag{7.38}$$

With this definition the propagator for a free particle can be calculated and is

$$K(x_1, p_0 T) = \exp((i/\hbar)(p_0 x_1 - (p_0^2/2m)T)) \tag{7.39}$$

Here we calculate the integral (7.38) using the SPA with the semi-classical form (7.19) for the propagator $K(x_1, x_0, T)$. In order to get the correct Morse index it would be better to use the more powerful methods of Möhring *et al.* (1980).

The saddle point in the integral (7.38) is determined by the condition

$$\partial S/\partial x_0 + p_0 = 0 \tag{7.40}$$

Comparing with (7.13) we see that the saddle point condition imposes boundary conditions

$$p(0) = p_0, \quad x(T) = x_1 \tag{7.41}$$

on the classical path used to calculate $S$. We introduce a modified action defined by a Legendre transformation

$$\begin{aligned}\bar{S}(x_1, p_0, T) &= S(x_1, x_0, T) + p_0 x_0 \\ &= S - x_0(\partial S/\partial x_0)\end{aligned}$$

It is the natural action function if $(x_1, p_0, T)$ are taken as independent variables. The relations analogous to (7.12)–(7.14) are

$$\partial \bar{S}/\partial x_1 = p_1, \quad \partial \bar{S}/\partial p_0 = x_0, \quad \partial \bar{S}/\partial T = -E \tag{7.42}$$

To complete the evaluation of the integral (7.38) we need to calculate the second derivative of $S$ with respect to $x$. It is from (7.13)

$$\partial^2 S/\partial x_0^2 = -\partial p_0/\partial x_0 \tag{7.43}$$

Combining (7.43) with the pre-exponential factor in (7.19) for $K(x_1, x_0, T)$ finally gives the result

$$K(x_1, p_0, T) = e^{-\frac{1}{2}i\pi\bar{\nu}}\left[\frac{\partial^2 \bar{S}}{\partial x_1 \partial p_0}\right]^{\frac{1}{2}} \exp((i/\hbar)\bar{S}(x_1, p_0, T)) \tag{7.44}$$

The index $\bar{\nu}$ in (7.44) is the same as the index $\nu$ in (7.19) if the second derivative in (7.43) is positive. Otherwise, there is an additional phase factor coming from the SPA result for the integral over $x_0$ (see appendix B). If $V(x) \to 0$ for $|x| \to \infty$ and $T$ is large enough so that the end points $(x_0, x_1)$ of the stationary path are both in regions where $V(x) = 0$ then the exponential factor has the magnitude

$$\frac{\partial^2 \bar{S}}{\partial x_1 \partial p_0} = \frac{\partial p_1}{\partial p_0} = 1$$

## 7.6 Relation with conventional quantum mechanics

Feynman's propagator (7.1) is a matrix element of the unitary time development operator in quantum mechanics

$$K(x_1, x_0, T) = \langle x_1 | \exp(-iHT/\hbar) | x_0 \rangle \tag{7.45}$$

Schulman (1981) uses (7.45) as a definition of the propagator in his book on *Techniques and Applications of Path Integration.* It is the starting point for his definition of the path integral. Feynman & Hibbs (1965) do not write (7.45) explicitly. Instead they use a 'rule for two successive events'

$$K(x_b, x_a, T_1 + T_2) = \int K(x_b, x_c, T_2) K(x_c, x_a, T_1) dx_c$$

This rule is a consequence of equation (7.45).

The semi-classical formula (7.19) is exact for free particle motion. The Lagrangian of a free particle is $\mathscr{L} = \frac{1}{2}m\dot{x}^2$ and the action is

$$S(x_1, x_0, T) = \tfrac{1}{2}m(x_1 - x_0)^2/T \tag{7.46}$$

Substituting (7.46) into (7.19) yields

$$K(x_1, x_0, T) = (m/2\pi i\hbar T)^{\frac{1}{2}} \exp(\tfrac{1}{2}im(x_1 - x_0)^2/\hbar T) \tag{7.47}$$

Equation (7.19) also gives an exact formula for the propagator in any potential $V(x)$ which is at most quadratic in $x$. Then equation (7.22) is exact and subsequent steps in the derivation of (7.19) involve no approximation. For example, if $V(x) = -\frac{1}{2}m\omega^2 x^2$ is the potential of a parabolic potential barrier then the classical action

$$\left.\begin{aligned} S(x_1, x_0, T) &= \frac{m\omega}{2\sinh(\omega T)}[(x_1^2 + x_0^2)\cosh(\omega T) - 2x_1 x_0] \\ \frac{i}{2\pi\hbar}\frac{\partial^2 S}{\partial x_1 \partial x_0} &= \left[\frac{m\omega}{2\pi i\hbar \sinh(\omega T)}\right] \end{aligned}\right\} \tag{7.48}$$

Substituting into (7.19) gives the exact propagator.

The amplitude (7.38) for propagating from an initial momentum $p_0$ to a final position $x_1$ is related to the unitary time development operator by

$$K(x_1, p_0, T) = \langle x_1 | \exp(-iHT/\hbar) | p_0 \rangle \tag{7.49}$$

For free particle motion $|p_0\rangle$ is an eigenstate of $H$ with eigenvalue $E_0 = \frac{1}{2}p_0^2/m$

$$\begin{aligned} K(x_1, p_0, T) &= \langle x_1 | p_0 \rangle \exp(-iE_0 T/\hbar) \\ &= \exp(ip_0 x_1/\hbar) \exp(-iE_0 T/\hbar) \end{aligned}$$

This formula is the same as (7.39).

Green's operator is defined by

$$\left.\begin{aligned} G^{+}(E) &= (1/i\hbar)\int_0^\infty dT \exp[-\varepsilon T/\hbar + i(E-H)T/\hbar] \\ &= \frac{1}{E + i\varepsilon - H} \quad \text{as } \varepsilon \to 0^+ \end{aligned}\right\} \tag{7.50}$$

Here $\varepsilon$ is a positive number which tends to zero. It gives a convergence factor in the integral (7.50). Comparing (7.50) and (7.31) we see that the propagator in the energy representation

$$G(x_1, x_0, E) = \langle x_1 | G^{+}(E) | x_0 \rangle$$

Suppose that the potential $V(x) \to 0$ as $x \to \pm\infty$ and that $\psi_+(x)$ and $\psi_-(x)$ are two solutions of the Schrödinger equation with boundary conditions

$$\begin{aligned} \psi_+(x) &\to A_+ \exp(ikx); \qquad x \to \infty \\ \psi_-(x) &\to A_- \exp(-ikx); \quad x \to -\infty \end{aligned}$$

where $A_+$ and $A_-$ are constants. The propagator satisfying outgoing wave boundary conditions can be expressed exactly in terms of $\psi_+$ and $\psi_-$ as

$$\left.\begin{aligned} G(x_1, x_0, E) &= B\psi_+(x_1)\psi_-(x_0); \quad x_1 > x_0 \\ &= B\psi_-(x_1)\psi_+(x_0); \quad x_1 < x_0 \end{aligned}\right\} \tag{7.51}$$

$$B = 2m/(\hbar^2 w(\psi_-, \psi_+))$$

where $w(\psi_-, \psi_+) = \psi_- \psi'_+ - \psi_+ \psi'_-$ is the Wronskian of $\psi_-$ and $\psi_+$ and is independent of $x$.

If WKB wave functions are used for $\psi_\pm(x)$ then (7.51) gives the semi-classical propagator. Its form depends on the nature of the potential and on the value of $E$. If $E > V(x)$ for all $x$ then

$$\psi_\pm(x) \approx (k(x))^{-\frac{1}{2}} \exp(\pm iW(x,b)/\hbar)$$

$$W(x,b) = \int_b^x p(x)dx \tag{7.52}$$

and $p(x) = \hbar k(x) = (2m(E - V(x))^{\frac{1}{2}}$. If the derivatives $\psi'_\pm(x)$ are calculated remembering equation (3.8) then the Wronskian $w = 2i/\hbar$ and (7.50) reduces to the semi-classical propagator (7.37).

As another example, suppose that $V(x)$ has a potential barrier as in fig. 7.1. We choose $\psi_+(x)$ to have an ingoing and reflected wave to the left of the barrier and an outgoing wave to the right

$$\begin{aligned} \psi_+(x) &\approx (k(x))^{-\frac{1}{2}}[\exp(iW(x,a)/\hbar) + R\exp(-iW(x,a)/\hbar)]; \quad x < a \\ &\approx (k(x))^{-\frac{1}{2}} T \exp(iW(x,b)/\hbar); \quad x > b \end{aligned}$$

and suppose that $\psi_-(x)$ has an outgoing wave to the left of the barrier

$$\psi_-(x) \approx (k(x))^{-\frac{1}{2}} \exp(-iW(x,a)/\hbar)$$

Again the Wronskian $w = 2i/\hbar$. If the points $x_0$ and $x_1$ are on opposite sides of the barrier ($x_0 < a, x_1 > b$) then equation (7.51) gives

$$G(x_1, x_0, E) = (1/i\hbar)(\dot{x}_0\dot{x}_1)^{-\frac{1}{2}} T \exp\{(i/\hbar)(W(x_1, b) + W(a, x_0))\} \quad (7.53)$$

The derivation of (7.53) uses the WKB wave functions near $x_0$ and $x_1$. No assumption is made about the wave functions in the barrier region. Equation (7.53) can be taken as a definition of the transmission amplitude $T$. In the next section $G(x_1, x_0, E)$ is calculated by the path integral method and $T$ can be extracted by comparison with (7.53).

## 7.7 Barrier penetration by the path integral method

A formula for the penetration coefficient of a potential barrier was obtained in appendix E by using a uniform approximation to solve the Schrödinger equation. The same problem can also be solved by calculating the time integral for the propagator $G(x_1, x_0, E)$ by a uniform approximation (Brink & Smilansky, 1983).

The potential is shown in fig. 7.1. The incident energy $E$ is less than the potential energy $V_B$ at the top of the barrier and we aim to calculate the propagator $G(x_1, x_0, E)$ from an initial point $x_0$ to the left of the barrier to a final point $x_1$ to the right of the barrier. The process is therefore forbidden in a classical theory. The points $x_0, x_1$, are taken to be far from the classical turning points at $b$ and $a$. The semi-classical expression for the propagator in time is given by (7.19)

$$K(x_1, x_0, T) = \left[\frac{i}{2\pi\hbar}\frac{\partial^2 S}{\partial x_0 \partial x_1}\right]^{\frac{1}{2}} \exp((i/\hbar)S(x_1, x_0, T)) \quad (7.54)$$

In the case of a parabolic barrier with potential

$$V = -\tfrac{1}{2}m\omega^2 x^2$$

the expression (7.54) for $K$ is exact with $S$ and $\partial^2 S/\partial x_0 \partial x_1$ given by (7.48). The parabolic barrier will be used as a guide to suggest a uniform approximation but the results obtained in this section hold for a more general barrier.

For a potential of the form shown in fig. 7.1 there is a unique real path satisfying the classical equations of motion from $x_0$ to $x_1$ for any real time $T$. The relation between the time and energy of the path is given by

$$T = \int_{x_0}^{x_1} (m/p(x))dx \quad (7.55)$$

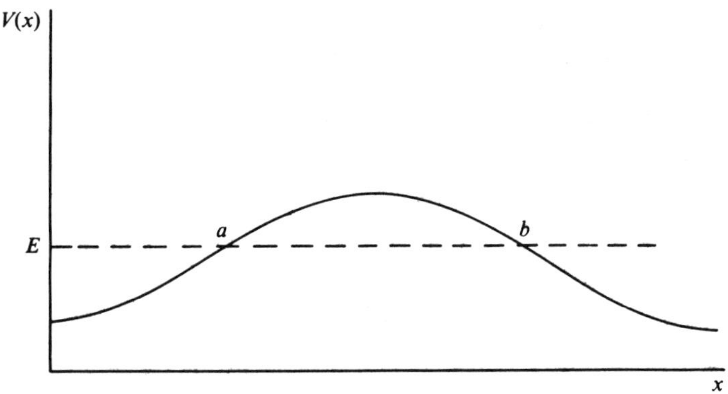


Fig 7.1. A potential barrier $V(x)$ with classical turning points $(V(x)=E)$ at $x=a$ and $x=b$.

The propagator $G(x_1, x_0, E)$ is given by the integral (7.31) and as shown in section 7.4 this can be calculated by the SPA. The stationary time $T$ is the time for a classical path from $x_0$ to $x_1$ with energy $E$. If $E > V_B$ there is one real stationary time and $G$ is given by (7.37). When $E < V_B$ there is no stationary point on the real axis but the integrand has saddle points in the complex $T$-plane. The parabolic barrier has a sequence of saddle points corresponding to complex times

$$T_s = T_0 - (2n+1)i\tau \tag{7.56}$$

where $n$ is an integer and $\tau = \pi/\omega$. $T_0$ is the sum of the classical times for propagation from $x_0$ to $a$, and from $b$ to $x_1$. There is a steepest descent path starting at $T=0$ and passing over the $n=0$ saddle in the lower half of the complex $T$ plane. In the case of the general barrier of fig. 7.1 the arrangement of saddles is the same as in (7.56) where

$$\left.\begin{aligned} \tau &= \int_a^b (m/\hbar\gamma(x))dx \\ \hbar\gamma(x) &= (2m(V(x)-E))^{\frac{1}{2}} \end{aligned}\right\} \tag{7.57}$$

is the complex time for propagation under the barrier. The saddle point contribution to the integral can be calculated just as in section 7.4 and gives the result

$$G(x_1, x_0, E) = (1/i\hbar)(\dot{x}_0\dot{x}_1)^{-\frac{1}{2}}\exp((i/\hbar)W(x_1, x_0, E)) \tag{7.58}$$

$$W(x_1, x_0, E) = W(x_1, b) + W(a, x_0) + iQ$$

$$Q = \int_a^b \gamma(x)dx = \pi(-E)/\hbar\omega$$

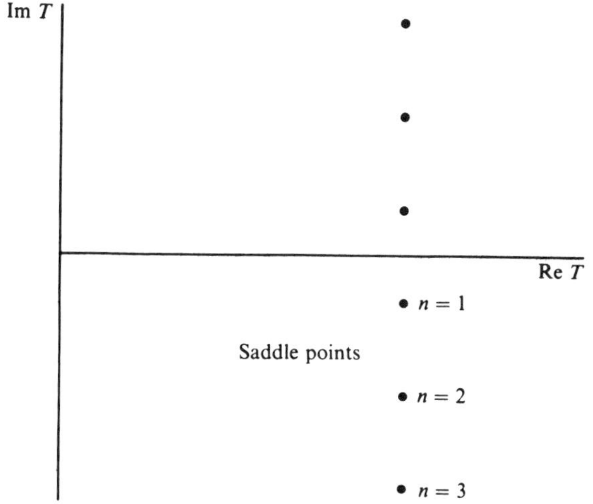


Fig. 7.2. The dots represent positions of the saddle points in equation (7.56) in the complex $T$-plane when $E < V_B$.

for the parabolic barrier. Comparing with (7.53) of section 7.6 the transmission amplitude is the same as the primitive WKB result (E.4). The derivation of the transmission amplitude by this method was first given by Freed (1972).

If the incident energy is near the barrier top the above result for $T$ is no longer valid. Various authors (Freed, 1972; Coleman, 1977) have suggested that a better result can be obtained by summing amplitudes corresponding to multiple reflections under the barrier. This corresponds to summing contributions from the line of saddles in the lower half $T$-plane in fig. 7.2. Such a procedure is unreliable for the following reason: When $E \to V_B$ the real part of the barrier position $T_0 \to \infty$ because the particle moves very slowly near the barrier. When $E = V_B$ the particle stops on top of the barrier and remains there. For the parabolic barrier

$$\omega T_0 \simeq \ln(2m\omega^2 |x_0 x_1/E|) \qquad (7.59)$$

The spacing between the saddle points in (7.56) is $2\pi/\omega$ and the saddle point width is $|\partial^2 S/\partial T^2|^{-\frac{1}{2}} \simeq (\hbar/\omega|E|)^{\frac{1}{2}}$ Hence the ratio

$$\text{width/spacing} \simeq (1/2\pi)(\hbar\omega/|E|)^{\frac{1}{2}} \qquad (7.60)$$

When $|E| \ll \hbar\omega$ the width of the saddles becomes comparable with or larger than their spacing and the approximation of summing the contributions of independent saddles is no longer valid. In this case the contribution of all the saddles can be summed by using a uniform approximation to evaluate the intergral for $G$. The logarithmic divergence (7.59) in Re $T$ and the relation (7.60) are characteristic of any barrier top.

In the case of a parabolic barrier when the incident energy is near the barrier top, $\hbar\omega \gg V_B - E$ then Re $T\omega \gg 1$ in the region of the complex plane containing the saddle points (7.56). The action function (7.54) can be approximated by

$$ET + S \simeq \tfrac{1}{2}m\omega[(x_0^2 + x_1^2) + 4|x_0 x_1|\exp(-\omega T)] - |E|T$$

The form of this expression suggests a comparison function for the uniform approximation to the integral (7.32) for $G$

$$\phi(\beta, y) = e^{-y} - \beta y \tag{7.61}$$

The stationary points of $ET + S$ are given by (7.56) while the stationary points of $\phi$ are

$$y_n = -\ln\beta - (2n+1)i\pi$$

In both cases there is an infinite sequence of stationary points. The mapping condition (C.6) at the stationary points is

$$(ET + S)_n = \phi(\beta, y_n) + A \tag{7.62}$$

where

$$(ET + S)_n = W(x_1, b) + W(a, x_0) + (2n+1)iQ$$
$$\phi(y_n, \beta) = \beta\ln\beta - \beta + (2n+1)i\pi\beta$$

The integer can be interpreted as the number of reflections under the barrier. The mapping condition (7.62) gives

$$\pi\beta = Q \tag{7.63}$$

and

$$A = W(x_1, b) + W(a, x_0) - (\beta\ln\beta - \beta) \tag{7.64}$$

The remaining term in the uniform integral (C.9) is the factor $h(y)$ containing the pre-exponential factor $g(T)$ and the jacobian of the transformation from $T$ to $y$. At the saddle points (see C.12)

$$h(y_n) = \exp(-\tfrac{1}{2}y_n)[2\pi i\dot{x}_0\dot{x}_1]^{-\frac{1}{2}} \tag{7.65}$$

Equation (7.65) holds exactly at the saddle point positions. It is natural to make the approximation that (7.65) holds for all values of $y$. Then (C.10) gives

$$G(x_1, x_0, E) = (1/i\hbar)[2\pi i\dot{x}_0\dot{x}_1]^{-\frac{1}{2}}\exp(iA)I_0(\beta) \tag{7.66}$$

$$I_0(\beta) = \int_{-\infty}^{\infty} \exp(ie^{-y} - (\tfrac{1}{2} + i\beta)y)dy \tag{7.67}$$

The integration limits $(0, \infty)$ for the $T$ integration in the integral (7.31) for G transform into the limits $(-\infty, \infty)$ for $y$. The integrand in (7.67) oscillates as $y \to \infty$ along the real axis but the integral converges if $y$ has a small negative imaginary part. Then the integral can be evaluated in terms of a Γ-function and gives the result

$$I_0(\beta) = \exp(\tfrac{1}{4}i\pi - \tfrac{1}{2}\pi\beta)\Gamma(\tfrac{1}{2} + i\beta)$$

The transmission amplitude obtained from (7.66)

$$T = (2\pi)^{-\frac{1}{2}}\Gamma(\tfrac{1}{2} + i\beta)\exp(-\tfrac{1}{2}\pi\beta - i(\beta\ln\beta - \beta)) \tag{7.68}$$

This is identical with (E.20) which was found in appendix $E$ by a uniform approximation to the Schrödinger equation. It reduces to the primitive WKB formula (7.58) if the energy is well below the top of the barrier ($\beta \gg 1$).

Equation (7.66) for $G$ is an analytical function of $E$ for $E$ near $V_B$. There is a logarithmic singularity in $(W(x_1, b) + W(x_0, a))$ which is cancelled by a corresponding singularity in $\beta\ln\beta$. Hence (7.66) is valid both above and below the barrier.

If the potential $V(x)$ is real then $\beta$ is real and the transmission probability is given by the well-known formula (Kemble, 1935)

$$\begin{aligned}|T^2| &= \exp(-\pi\beta)|\Gamma(\tfrac{1}{2} + i\beta)|^2/2\pi \\ &= 1/(1 + e^{2\pi\beta})\end{aligned} \tag{7.69}$$

In the case of a parabolic barrier

$$\beta = (V_B - E)/\hbar\omega$$

# 8

# Elastic scattering by the path integral method

## 8.1 A path integral formula for the scattering wave function

A semi-classical theory of elastic scattering based on the partial wave expansion of the scattering amplitude was developed in chapters 3 and 4. The scattering phase shifts were approximated by the WKB formula (3.49) and the partial wave sum was transformed into a sum of integrals (4.15) by using the Poisson sum formula. These integrals were evaluated by the stationary phase approximation to give a semi-classical expression (4.30) for the scattering amplitude. In the present chapter we give an alternative approach based on the path integral method. The first step is to develop a path integral formula for the scattering wave function. Then the path integral is evaluated by the stationary phase approximation. The result is the same as equation (4.30). The method is valuable because it can also be used to study more complicated reactions.

Scattering problems are usually formulated in terms of the scattering wave function which satisfies the Schrödinger equation and the boundary condition

$$\psi(\mathbf{r}) \sim e^{ikz} + \frac{1}{r} e^{ikr} f(\theta) \tag{8.1}$$

when $r$ is large. In (8.1) the incident wave is parallel to the $z$-axis, $\theta$ is the scattering angle and the incident energy is $E = \hbar^2 k^2/2m$. Feynman's propagator $K(\mathbf{r}, \mathbf{p}_0, T)$, which gives the amplitude for a final position $\mathbf{r}$ at time $T$ when the initial momentum is $\mathbf{p}_0$, is related to the scattering wave function and we shall show that

$$\psi(r) = \lim_{T \to \infty} \bar{K}(\mathbf{r}, \mathbf{p}_0, T) \exp(iE_0 T/\hbar) \tag{8.2}$$

where $\mathbf{p}_0$ is parallel to the $z$-axis and $E = E_0 = \mathbf{p}_0^2/2m$. The propagator $K$ is the three-dimensional generalization of the one discussed in section 7.5.

The connection between the path integral method and conventional quantum mechanics was discussed in section 7.6. Propagators are matrix elements of the unitary time development operator and

$$\bar{K}(\mathbf{r}, \mathbf{p}_0, T) = \langle \mathbf{r} | \exp(-iHT/\hbar) | \mathbf{p}_0 \rangle \tag{8.3}$$

where $|\mathbf{r}\rangle$ and $|\mathbf{p}_0\rangle$ are states in the coordinate and momentum representations. $H = H_0 + V$ is the hamiltonian, $H_0$ is the kinetic energy operator and $|\mathbf{p}_0\rangle$ is an eigenstate of $H_0$

$$H_0|\mathbf{p}_0\rangle = (\mathbf{p}_0^2/2m)|\mathbf{p}_0\rangle = E_0|\mathbf{p}_0\rangle \tag{8.4}$$

The propagators $\bar{K}$ and $K$ are related by a generalization of equation (7.38)

$$\bar{K}(\mathbf{r}, \mathbf{p}_0, T) = \int d\mathbf{r}_0 \exp(i\mathbf{r}_0 \cdot \mathbf{p}_0/\hbar) K(\mathbf{r}, \mathbf{r}_0, T) \tag{8.5}$$

The scattering wave function is given by a matrix element of the Møller wave operator (Taylor, 1972)

$$\psi(\mathbf{r}, \mathbf{p}_0) = \langle \mathbf{r}|\Omega_+|\mathbf{p}_0\rangle \tag{8.6}$$

$$\Omega_+ = \lim_{T\to\infty} \exp(-iHT/\hbar)\exp(iH_0T/\hbar) \tag{8.7}$$

Equation (8.2) for the scattering wave function is obtained by substituting the operator (8.7) into the matrix element (8.6) and then using equations (8.4) and (8.3).

This derivation has to be modified in the presence of a Coulomb potential because of the long-range character of the field. If the incident wave has wave number $k$ and is travelling in the direction of the positive $z$-axis the asymptotic form (8.1) of the scattering wave function becomes

$$\psi(\mathbf{r}) \sim e^{i(kz + n\ln k(r-z))} + \frac{1}{r}f(\theta)e^{i(kr - n\ln 2kr)} \tag{8.8}$$

For this case the limit in (8.7) does not exist and it is not possible to define the usual Møller operator. However, Dollard (1964) shows that the time-independent wave function $\psi(\mathbf{r})$ can be obtained from equation (8.6) using a modified Møller operator

$$\Omega_{C+} = \lim_{T\to\infty} \exp(-iHT/\hbar)\exp(iH_0T/\hbar + iH_C(T)/\hbar) \tag{8.9}$$

$$H_C(T) = \hbar n \ln(4H_0T/\hbar) \tag{8.10}$$

With these modifications equation (8.2) for the scattering wave function is replaced by

$$\psi(\mathbf{r}, \mathbf{p}_0) = \lim_{T\to\infty} \bar{K}(\mathbf{r}, \mathbf{p}_0, T)\exp\{i\phi_C(T)\} \tag{8.11}$$

$$\phi_C(T) = E_0T/\hbar + n\ln(4E_0T/\hbar) \tag{8.12}$$

and has the correct asymptotic form (8.8) for a wave function in the presence of a Coulomb field.

Koeling & Malfliet (1975) and Miller (1970, 1974) have studied a path integral formulation of scattering theory based on the $S$-matrix rather than the Møller wave operator. They relate the $S$-matrix to the momentum space propagator $K(\mathbf{p}_1, \mathbf{p}_0, T)$ where $\mathbf{p}_0$ is the momentum at $t=0$ and $\mathbf{p}_1$ is the momentum at $t=T$.

## 8.2 The semi-classical wave function

We consider the scattering of a particle with mass $m$ by a potential $V(r)$. The propagator $\bar{K}$ in equation (8.2) can be written as a path integral

$$\bar{K}(\mathbf{r}, \mathbf{p}_0, T) = \int D[\mathbf{r}(t)] \exp((i/\hbar)\bar{S}[\mathbf{r}(t)]) \tag{8.13}$$

where

$$\begin{aligned}\bar{S}[\mathbf{r}(t)] &= S[\mathbf{r}(t)] + \mathbf{p}_0 \cdot \mathbf{r}_0 \\ &= \int_0^T \mathscr{L}(\mathbf{r}, \dot{\mathbf{r}})dt + \mathbf{p}_0 \cdot \mathbf{r}_0\end{aligned} \tag{8.14}$$

where $\mathscr{L} = \frac{1}{2}m\dot{\mathbf{r}}^2 - V(\mathbf{r})$ is the Lagrangian. The path integral in (8.13) is taken over all paths satisfying the boundary conditions

$$m\dot{\mathbf{r}}(0) = \mathbf{p}_0, \quad \mathbf{r}(T) = \mathbf{r} \tag{8.15}$$

A semi-classical expression for the propagator is obtained by evaluating (8.13) by the stationary phase approximation. A stationary path satisfies the condition

$$\left.\begin{aligned}\delta\bar{S}[\mathbf{r}(t)] &= \int_0^T \left(\frac{\partial \mathscr{L}}{\partial \mathbf{r}} - \frac{d}{dt}\left(\frac{\partial \mathscr{L}}{\partial \dot{\mathbf{r}}}\right)\right) \cdot \delta\mathbf{r} dt + \mathbf{p} \cdot \delta\mathbf{r} + \mathbf{r}_0 \cdot \delta\mathbf{p}_0 \\ &= 0\end{aligned}\right\} \tag{8.16}$$

for small variations $\delta\mathbf{r}(t)$ about a path $\mathbf{r}(t)$. Paths determined by (8.16) satisfy the classical equations of motion with boundary conditions (8.15). The action $\bar{S}(\mathbf{r}, \mathbf{p}_0, T)$ for a stationary path is a function of the final position $\mathbf{r}$, the initial momentum $\mathbf{p}_0$ and the time interval $T$. Equation (8.16) shows that

$$\partial\bar{S}/\partial\mathbf{r} = \mathbf{p} \quad \text{and} \quad \partial\bar{S}/\partial\mathbf{p}_0 = \mathbf{r}_0 \tag{8.17}$$

These relations are natural generalizations of (7.42) to several dimensions.

The stationary phase evaluation of the path integral gives a result which is a natural generalization of (7.44)

$$\bar{K}(\mathbf{r}, \mathbf{p}_0, T) = \sum_\alpha A_\alpha(\mathbf{r}, \mathbf{p}_0) \exp((i/\hbar)\bar{S}_\alpha(\mathbf{r}, \mathbf{p}_0, T)) \tag{8.18}$$

where the sum is taken over all stationary paths satisfying the boundary conditions (8.15).

The amplitude

$$A_\alpha(\mathbf{r}, \mathbf{p}_0) = |D_\alpha(\bar{S})|^{\frac{1}{2}} \exp(-\tfrac{1}{2}i\pi\nu_\alpha) \tag{8.19}$$

where $\nu_\alpha$ is a Maslov index. The van Vleck determinant $D_\alpha(\bar{S})$ can be written in three equivalent ways

$$D(\bar{S}) = \det\left(\frac{\partial^2 \bar{S}}{\partial r_i \partial p_{0j}}\right) = \det\left(\frac{\partial p_i}{\partial p_{0j}}\right) = \det\left(\frac{\partial r_{0j}}{\partial r_i}\right) \tag{8.20}$$

Equation (8.19) for the pre-exponential factor $A_\alpha$ can be derived by a generalization of the method discussed in section 7.1. This derivation due to Levit & Smilansky (1977*a*) gives the phase as well as the magnitude of $A_\alpha$ but it is too long to reproduce here. Instead, we give a very simple unitary argument due to Miller (1970) which leads to the correct magnitude for $A_\alpha$, but does not give the phase. Miller's argument can only be used when there is just one term in formula (8.18). There are complications when there are several terms because of interference effects. We suppress the time interval $T$ because it is constant throughout the calculations.

Unitarity gives

$$\begin{aligned}
(2\pi\hbar)^3 \delta(\mathbf{r}-\mathbf{r}') &= \int \bar{K}(\mathbf{r},\mathbf{p}_0)\bar{K}^*(\mathbf{r}',\mathbf{p}_0)d\mathbf{p}_0 \\
&= \int d\mathbf{p}_0 A(\mathbf{r},\mathbf{p}_0)A^*(\mathbf{r}',\mathbf{p}_0)\exp\left[(i/\hbar)(\bar{S}(\mathbf{r},\mathbf{p}_0)-\bar{S}(\mathbf{r}',\mathbf{p}_0))\right] \\
&\simeq \int d\mathbf{p}_0 |A(\mathbf{r},\mathbf{p}_0)|^2 \exp\left[(i/\hbar)(\mathbf{r}-\mathbf{r}')\cdot\mathbf{p}(\mathbf{r},\mathbf{p}_0)\right]
\end{aligned}$$

In the last step $A(\mathbf{r},\mathbf{p}_0)$ is assumed to be a slowly varying function of $\mathbf{r}$ and $\bar{S}(\mathbf{r}',\mathbf{p}_0)$ is expanded in a Taylor series about $\mathbf{r}$ to first order in $(\mathbf{r}'-\mathbf{r})$. From equation (8.17) $\mathbf{p}(\mathbf{r},\mathbf{p}_0)$ is the momentum at the final position $\mathbf{r}$. Changing the integration variable from $\mathbf{p}_0$ to $\mathbf{p}$ gives

$$(2\pi\hbar)^3\delta(\mathbf{r}-\mathbf{r}') = \int d\mathbf{p}|A|^2 \det\left(\frac{\partial p_{0i}}{\partial p_j}\right)\exp\left[(i/\hbar)(\mathbf{r}-\mathbf{r}')\cdot\mathbf{p}\right] \tag{8.21}$$

where the determinant is the jacobian of the transformation. Equation (8.21) is satisfied provided the product of $|A|^2$ with the jacobian is unity. Hence

$$|A(\mathbf{r},\mathbf{p}_0)|^2 = |\det(\partial p_j/\partial p_{0i})|$$

which is consistent with equation (8.19).

A semi-classical formula for the scattering wave function is obtained by substituting (8.18) into (8.11)

$$\psi(\mathbf{r},\mathbf{p}_0) = \sum_\alpha A_\alpha(\mathbf{r},\mathbf{p}_0)\exp(iW_\alpha(\mathbf{r},\mathbf{p}_0)/\hbar) \tag{8.22}$$

where

$$W_\alpha(\mathbf{r},\mathbf{p}_0) = \lim_{T\to\infty}\left[\bar{S}_\alpha(\mathbf{r},\mathbf{p}_0,T)+\phi_C(E_0,T)\right]$$

The action $\bar{S}_\alpha$ is given by equation (8.14). It can also be written as

$$\bar{S}(\mathbf{r}, \mathbf{p}_0, T) = \int_{\mathbf{r}_0}^{\mathbf{r}} \mathbf{p}\cdot d\mathbf{r} - ET + \mathbf{p}_0\cdot\mathbf{r}_0$$

where the integral is taken along the classical path $\mathbf{r}(t)$ with final position $\mathbf{r}$ and initial momentum $\mathbf{p}_0$. When $T$ is large the position $\mathbf{r}_0$ at $t = 0$ is a point on the trajectory a long time before the scattering. Now we show that $W_\alpha$ can be written in the form

$$W_\alpha(\mathbf{r}, \mathbf{p}_0) = \int_{\mathbf{r}_0}^{\mathbf{r}} \mathbf{p}\cdot d\mathbf{r} - \hbar(kr_0 - n\ln(2kr_0)) + X \tag{8.23}$$

where $X \to 0$ as $T \to \infty$. The expression for $X$ is obtained from (8.22) by writing $W$ in terms of $\bar{S}$ and $\phi_C$ and is

$$\begin{aligned} X &= \hbar(kr_0 - n\ln(2kr_0)) - ET + \mathbf{p}_0\cdot\mathbf{r}_0 + \phi_C(E_0, T) \\ &= \hbar kr_0 + \mathbf{p}_0\cdot\mathbf{r}_0 + (E_0 - E)T + \hbar n\{\ln(4E_0T/\hbar) - \ln(2kr_0)\} \end{aligned}$$

The total energy $E$ of the trajectory is constant throughout the motion and $E_0 = p_0^2/2m$ is the kinetic energy at $t = 0$. These differ by the Coulomb potential at $r_0$,

$$E - E_0 = V_C(r_0)$$

The difference is small and tends to zero as $T \to \infty$ but it is multiplied by $T$ in the expression for $X$. In the same way

$$|\mathbf{p}_0| \approx \hbar k(1 - V_C(r_0)/E)^{\frac{1}{2}} = \hbar k - V_C(r_0)/v$$

where $v$ is the incident speed of the projectile. Hence

$$\mathbf{r}_0\cdot\mathbf{p}_0 = -r_0p_0 \approx -\hbar kr_0 + r_0V_C(r_0)/v$$

Collecting these expression for $X$ we get

$$X = r_0V_C(r_0)/v - TV_C(r_0) + \hbar n\{\ln(4E_0T/\hbar) - \ln(2kr_0)\}$$

When $T$ is sufficiently large $T \simeq (r + r_0)/v$ and

$$X \simeq -rV_C(r_0)/v + \hbar n\ln(2k(1 + r/r_0))$$

As $T \to \infty$, $r/r_0 \to 0$ and $X \to 0$.

We conclude this section by discussing the physical meaning of equation (8.22). The square of the magnitude of the wave function is the probability density $P(\mathbf{r}, \mathbf{p}_0)$ for finding the scattered particle at $\mathbf{r}$. The probability density in the incident beam is unity and the incident momentum is $\mathbf{p}_0$. According to equation (8.22)

$$\begin{aligned} P(\mathbf{r}, \mathbf{p}_0) &= |\psi(\mathbf{r}, \mathbf{p}_0)|^2 \\ &= \sum_\alpha |A_\alpha(\mathbf{r}, \mathbf{p}_0)|^2 + \text{interference terms} \end{aligned} \tag{8.24}$$

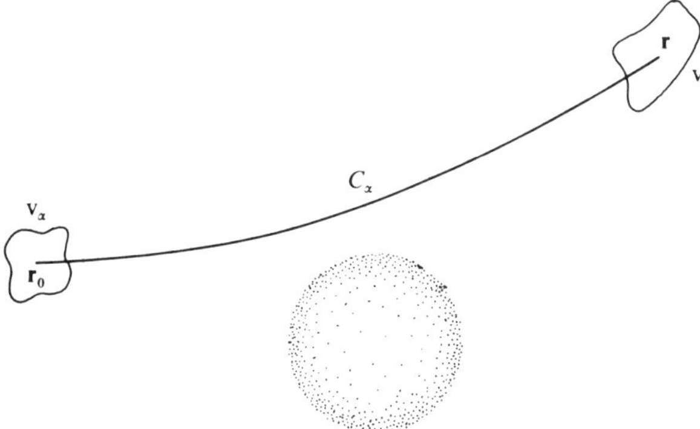


Fig. 8.1. Sketch of a classical orbit $C_\alpha$. Particles in a volume element $v$ near $\mathbf{r}$ at $t = T$ occupied a volume element $v_\alpha$ near $r_0$ at $t = 0$.

Now we consider the probability density for classical scattering. Particles in a volume element $v$ near $\mathbf{r}$ (fig. 8.1) which arrived there by travelling along a classical trajectory $\alpha$ occupied a volume $v_\alpha$ at an earlier time before the scattering. The probability density before the scattering was unity; so after it becomes

$$v_\alpha/v = |\det(\partial r_{0j}/\partial r_i)|_\alpha = |A_\alpha(\mathbf{r}, \mathbf{p}_0)|^2 \tag{8.25}$$

Equation (8.25) holds because the ratio of volume elements is just the jacobian of the transformation of coordinates produced by the equations of motion. If several trajectories pass through $\mathbf{r}$ then the total probability density is the sum of contributions like (8.25)

$$P_{\mathrm{cL}}(\mathbf{r}, \mathbf{p}_0) = \sum_\alpha |A_\alpha(\mathbf{r}, \mathbf{p}_0)|^2 \tag{8.26}$$

Thus the probability density predicted by the semi-classical theory is the same as for classical scattering with the addition of interference terms. This is consistent with results obtained in chapter 4.

Caustics are points, curves or surfaces where $A_\alpha$ given by (8.25) diverges and the classical cross-section (8.26) becomes infinite. The stationary phase formula (8.22) for the wave function breaks down near caustics, but it can be replaced by a uniform approximation (see section 8.6).

## 8.3 WKB in three dimensions

Equation (8.22) for the wave function can also be obtained by a conventional WKB approach (Messiah, 1959). We seek a solution of the

Schrödinger equation of the form

$$\psi(\mathbf{r}) = A(\mathbf{r}) \exp(iW(\mathbf{r})/\hbar) \tag{8.27}$$

This wave function is substituted into the Schrödinger equation and terms of order $\hbar^2$ are neglected. Collecting terms independent of $\hbar$ gives the Hamilton–Jacobi equation

$$(\nabla W)^2/2m + V(r) = E \tag{8.28}$$

and terms of order $\hbar$ give the continuity equation

$$\operatorname{div}(A^2 \nabla W) = 0 \tag{8.29}$$

The action function $W(\mathbf{r}, \mathbf{p}_0)$ in (8.26) automatically satisfies the Hamilton–Jacobi equation with energy $E = \mathbf{p}_0^2/2m$. This is because $\partial W/\partial \mathbf{r} = \mathbf{p}$ and (8.28) is just the energy conservation condition.

$W$ defined by (8.26) also has appropriate boundary conditions for scattering of particles with incident momentum $\mathbf{p}_0$. In equation (8.29) $|A|^2$ is the classical probability density. The interpretation of the equation is that probability flows along the classical trajectories and is conserved. Equation (8.29) can be solved (van Vleck, 1928) and the solution is equivalent to (8.25). The surfaces $|A|^2 = \infty$ are caustics of the classical scattering problem and are envelopes of the family of classical trajectories with incident momentum $\mathbf{p}_0$. The WKB solution (8.27) breaks down near the caustics as does the equivalent path integral wave function (8.22). The Maslov indices can be obtained from connection formulae which are generalizations of those in section 3.3 to several dimensions (van Horn & Salpeter, 1967; Schiller, 1962).

## 8.4 The scattering amplitude

In this section we evaluate the scattering amplitude for a spherical potential from the semi-classical wave function (8.22). We choose the incident momentum to be parallel to the $z$-axis and suppose that the point of observation $r$ is far from the scatterer. We also choose $T$ so large that $r_0 \gg r$. Fig. 8.2 shows two classical trajectories which contribute to the semi-classical wave function at $r$. The trajectory $C_0$ passes far from the scatterer and is undeflected. Evaluating the action integral (8.23) shows that $C_0$ contributes a term $\exp i(kz + n \ln k(r - z))$ to the wave function and corresponds to the incident wave in the asymptotic form (8.8) of the exact wave function.

The trajectory $C_1$ passes close to the target. It is convenient to evaluate

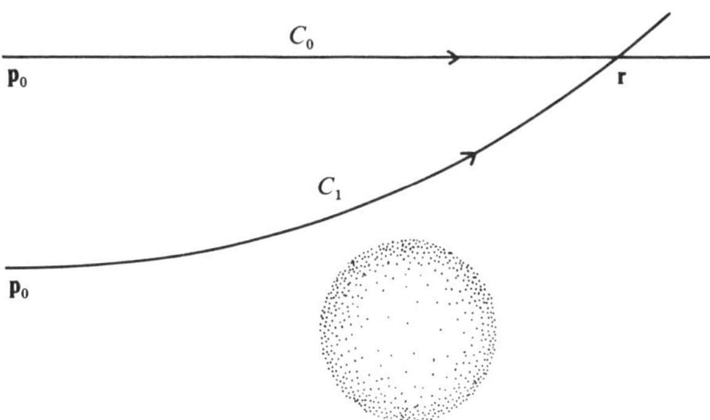


Fig. 8.2. $C_0$ and $C_1$ represent two orbits with the same initial momentum $\mathbf{p}_0$ and the same final position $\mathbf{r}$.

the action integral in (8.23) in polar coordinates

$$\int_{r_0}^{r} \mathbf{p}\cdot d\mathbf{r} = \int_{r_t}^{r} p_r dr + \int_{r_t}^{r_0} p_r dr + \int_{\pi}^{\theta(\lambda)} p_\theta d\theta \tag{8.30}$$

Here $p_r$ and $p_\theta$ are the radial and angular momenta and $r_t$ is the radial turning point,

$$p_r = 2m(E - V_{eff}(r))^{\frac{1}{2}}, \quad p_\theta = -\hbar\lambda$$

and $\Theta(\lambda)$ is the classical deflection angle. The effective potential $V_{eff}(r)$ is given in equation (3.50). The first two integrals in (8.30) are of the kind evaluated in section 3.6 and

$$\int_{r_t}^{r} p_r \, dr \sim \hbar(kr - n \ln 2kr + \delta(\lambda) - \tfrac{1}{2}\pi\lambda) \tag{8.31}$$

where $\delta(\lambda)$ is the semi-classical phase shift. The angular integral is

$$\int_{\pi}^{\Theta(\lambda)} p_\theta \, d\theta = -\hbar\lambda(\Theta(\theta) - \pi) \tag{8.32}$$

because $\lambda$ is constant along the orbit. Substituting these integrals into (8.23) gives

$$W(\mathbf{r}, \mathbf{p}_0)/\hbar \sim kr - n \ln 2kr + 2\delta(\lambda) - \lambda\Theta(\lambda) \tag{8.33}$$

Notice that the dependence on $r_0$ in (8.23) cancels out. In a classical problem the deflection angle can lie in the range $-\infty < \Theta(\lambda) < \pi$. As discussed in section 3.8 the observation angle $\theta$ is restricted to the range $0 < \theta < \pi$ and is related to the deflection angle by (3.54). Using this result $W$ can also be written as

$$W(\mathbf{r}, \mathbf{p}_0)/\hbar \sim kr - n \ln 2kr + (2\delta(\lambda) \mp \lambda\theta + 2m\pi\lambda) \tag{8.34}$$

The magnitudes of the pre-exponential factor in the semi-classical wave function (8.22) can be calculated using equation (8.25). Points in volume $v_1$ before the scattering occupy a volume $v$ after the scattering and $|A|^2 = v_1/v$. It is convenient to specify the initial volume $v_1$ in cylindrical polar coordinates $(z, b, \phi)$ where $b$ is the impact parameter. The volume element is

$$v_1 = bdbdzd\phi$$

The final volume is specified in polar coordinates $(r, \theta, \phi)$ and

$$v = r^2 dr \sin\theta d\theta d\phi$$

One has $dr = dz$ because the magnitude of the initial and final momenta are equal for elastic scattering. Combining these expression gives

$$\begin{aligned} |A_1|^2 &= v_1/v = (bdb/d\theta)/(r^2 \sin\theta) \\ &= \sigma_{\text{cL}}(\theta)/r^2 \end{aligned} \tag{8.35}$$

where $\sigma_{\text{cL}}$ is the classical cross-section. Substituting equations (8.34) and (8.35) into equation (8.22) for the wave function we notice that the scattered wave function has the correct $r$-dependence. If there are several classical paths with the same observation angle $\theta$ $(\Theta(\lambda_\alpha) = \pm\theta - 2m\pi)$ then the total scattering amplitude is a sum of contributions from each of these paths

$$f(\theta) = \sum_\alpha (\sigma_{\text{cL}}(\lambda_\alpha))^{\frac{1}{2}} \exp i(2\delta(\lambda_\alpha) - \lambda_\alpha \Theta(\lambda_\alpha) - \tfrac{1}{2}\pi v_\alpha) \tag{8.36}$$

The remaining ingredient in equation (8.36) for the semi-classical scattering amplitude is the Maslov index $v_\alpha$ which comes from (8.19) for $A_\alpha$. We give a prescription for calculating $v_\alpha$ based on Morse (1934) theory. The semi-classical amplitude (8.36) is determined by properties of a classical trajectory through $\mathbf{r}$ with incident momentum $\mathbf{p}_0$. The trajectory starts at $\mathbf{r}_0$ at $t = 0$ and arrives at $\mathbf{r}$ at $t = T$. Suppose that $T$ is large enough so that $r_0 \gg r$. Let $\mathbf{r}(t)$ be the position on the trajectory at $t$ and consider the determinant

$$D(t) = \det(\partial r_i(t)/\partial r_{0j}) \tag{8.37}$$

A point on the trajectory where $D(t)$ vanishes is called a focal point and its multiplicity is the number of zero eigenvalues of the matrix in (8.37) at that point. Fig. 8.3 shows some trajectories with the same incident momentum and slightly different impact parameters. In fig. 8.3($a$) there is no focal point and in fig. 8.3($b$) there is one where the orbits cross. In figs. 8.3($c$) and 8.3($d$) there are focal points on the $z$-axis. Orbits with the same impact

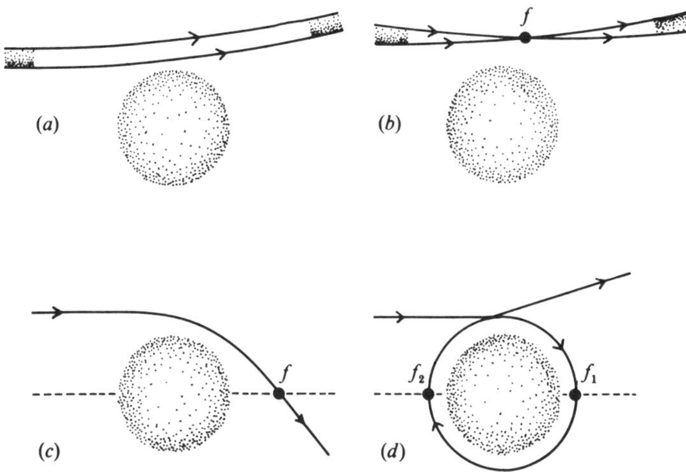


Fig. 8.3. Illustration of focal points. (*a*) no focal point, (*b*) focal point at *f*, (*c*) focal point at $\theta = 0$, (*d*) focal points at $\theta = 0$ and $\theta = \pi$.

parameter but different azimuthal angles $\phi$ cross there. The shaded areas in figs. 8.3(*a*) and 8.3(*b*) indicate the volume elements $v$ and $v_1$, which appear in equation (8.35). If a sign is associated with a volume element then the ratio $v_1/v$ is positive in fig. 8.3(*a*) and negative in fig. 8.3(*b*). Hence the phase of the semi-classical wave functions which are proportional to $(v_1/v)^{\frac{1}{2}}$ should change by $\pm i$ and the index $v_\alpha$ should change by $\pm 1$.

The conjecture based on Morse theory is that the change is always $+1$ and the index is equal to the number of the focal points on the path provided that each focal point is counted with its multiplicity. Levit *et al.* (1978) have shown that this prescription is valid for arbitrary initial conditions provided that the final condition fixes the coordinates. Their result is general enough to cover our case. It is simple to check that with the above method for choosing $v_\alpha$ equation (8.36) reproduces the results of chapter 4 in equation (4.30) for the cases illustrated in fig. 8.3.

## 8.5 Classically forbidden transitions and complex paths

In certain cases no real solutions of the classical equations can be found which satisfy the boundary conditions (8.15). In such cases a semi-classical approximation can be found by removing the requirement that classical trajectories should be real valued. This is done by the analytic continuation of the classical equations of motion into the complex domain, and the procedure is clearly analogous to replacing the sationary phase method by the more general saddle point approximation (appendix B) for evaluating the path integral. This complex extension is required for a discussion of rainbow scattering of heavy ions. It is also necessary for the study of the quantal effects of absorption by a complex optical potential. Various authors have discussed classically forbidden transitions by extending the classical equations of motion to the complex domain (Miller, 1974, 1975; Marcus, 1971; Knoll & Schaeffer, 1976, 1977). The ideas are almost the same as the extension to complex paths discussed in section 4.5. There is one important difference. The complex paths discussed in chapter 4 enter as a result of the saddle point evaluation of a one-dimensional integral over angular momentum. In the present chapter we consider a saddle point evaluation of a multi-dimensional path integral. The method is straightforward, namely complex paths have to be included in the summation in equation (8.22) for the scattering wave function. The problem lies in the fact that not all complex classical paths satisfying the boundary conditions (8.15) contribute. Only a restricted set should be retained and we need a criterion for deciding which are the 'good paths'. There is no complete answer to this question within the framework of path integrals. For central potentials the methods of chapter 4 give definite answers and the same paths can be used in equation (8.22). Then the two methods give identical results.

## 8.6 A diffraction formula

The stationary phase formula (8.22) for the scattering wave function breaks down in the neighbourhood of caustics where $A_\alpha(\mathbf{r}, \mathbf{p}_0)$ becomes infinite. In these circumstances it is possible to obtain good results by using a uniform approximation to evaluate the path integral.

Uniform approximations to finite dimensional integrals are discussed in appendix C. An integral $J$ is a uniform approximation to an integral $I$ if the stationary phase evaluation of both integrals gives the same result and if they have the same caustic structure. The extension to path integrals is possible because the method depends on the type of caustic and not on the number of dimensions. The crucial point is to determine a mapping function properly. Experience has shown that the number of caustic types, which occur in nuclear scattering problems, is very limited. Many of the caustics allowed by the general theory (Thom, 1975) do not occur. The reason for this is that most nuclear scattering processes have approximate cylindrical symmetry about an axis in the direction of the incident beam. We discuss a representation of the scattering amplitude by a diffraction integral which is general enough to cover the common caustics.

We begin by noting that there is no unique uniform approximation for a scattering amplitude. Different approximations should be asymptotically equal as $\hbar \to 0$. The integral (4.20) can be regarded as a uniform approximation to the path integral formula for the scattering amplitude. Evaluating it by the stationary phase approximation gives (4.30) and we have shown that equation (4.30) is identical to (8.22), which was obtained from the path integral (8.13). However, the formula (4.20) for the scattering amplitude is still singular at $\theta = 0$ and $\theta = \pi$, and it does not provide a good uniform approximation for the orbiting caustics in the forward and backward directions.

Now we give a diffraction integral which is a useful uniform approximation to $f(\theta)$ for forward and intermediate scattering angles. The formula is suggested by the results obtained with the Poisson sum formula in section 4.2. It can be written as an integral over a two-dimensional angular momentum vector $\lambda = k\mathbf{b}$ which is proportional to the impact parameter vector $\mathbf{b}$ and is

$$f(\theta) = \frac{1}{2\pi ik}\left(\frac{\theta}{\sin\theta}\right)^{\frac{1}{2}} \sum_{m=-\infty}^{\infty} (-1)^m \int_0^\infty \int_0^{2\pi} \lambda d\lambda d\phi \exp\{iF_m(\lambda, \phi, \theta)\} \quad (8.38)$$

where $\lambda$ and $\phi$ are the components of $\boldsymbol{\lambda}$ in cylindrical coordinates and

$$F_m(\lambda, \phi, \theta) = 2\delta(\lambda) - \lambda\theta\cos\phi + 2m\pi\lambda \quad (8.39)$$

It is a simple matter to check that the stationary phase evaluation of (8.38) gives the same result as the stationary phase formula (8.36) obtained from the path integral. Different $m$ values in (8.38) allow for deflection angles $\Theta(\lambda) = \pm\theta - 2m\pi$ which orbit around the scatterer several times before emerging as discussed in section 3.8. The $m = 0$ term gives the dominant contribution to $f(\theta)$ in most applications. The integral over $\phi$ in (8.38) can be calculated to give a Bessel function $J_0(\lambda\theta)$ and the resulting formula is equivalent to the Poisson sum formula for the scattering amplitude with the Legendre polynomials replaced by the Bessel function asymptotic form (4.18).

Equation (8.38) is a valid uniform approximation for forward scattering and for a rainbow caustic, but it is singular when $\theta = \pi$ and cannot be used for backward scattering. An alternative form, which is valid for the backward glory, is best written in terms of $\vartheta = \pi - \theta$,

$$f(\theta) = f(\pi - \vartheta)$$
$$= \frac{1}{2\pi i k}\left(\frac{\vartheta}{\sin\vartheta}\right)^{\frac{1}{2}} \sum_{n=-\infty}^{\infty} (-1)^n \int_0^{\infty}\int_0^{2\pi} \lambda d\lambda d\phi \exp\{i\bar{F}_n(\lambda, \phi, \vartheta)\} \qquad (8.40)$$

$$\bar{F}_n(\lambda, \phi, \vartheta) = 2\delta(\lambda) - \lambda\vartheta\cos\phi + (2n-1)\pi\lambda \qquad (8.41)$$

and is related to the Poisson sum formula (5.45) for $f(\theta)$ with the Bessel function asymptotic form (4.18) for the Legendre polynomials.

There is nothing new in equations (8.38) and (8.40). They are completely equivalent to the expressions for $f(\theta)$ obtained from the Poisson sum formula in chapters 4 and 5 and were suggested by them. But they are valid uniform approximations to the path integral for the elastic scattering amplitude and can be generalized to apply to inelastic scattering. This will be done in chapter 9.

Crowley (1978, 1980) has obtained another uniform approximation to the scattering amplitude by unfolding the singularity associated with forward and backward caustics. His formula is different from (8.38) and (8.40) but it is asymptotically equivalent to them when $\hbar \to 0$.

# 9
# Inelastic scattering

## 9.1 The scattering amplitude

In previous chapters of this book the various semi-classical theories of elastic scattering of nuclei were studied. Inelastic scattering and reactions are more complicated processes and it is impossible to review all the methods used in one chapter. We shall concentrate on a description based on the path integral method which gives a simple physical picture and which helps to place other methods in perspective. First, we define the scattering amplitude and other quantities needed for the discussion.

In general, three kinds of effects can be observed in a nucleus–nucleus collision. The first is elastic scattering

$$\mathrm{A} + \mathrm{X} \rightarrow \mathrm{A} + \mathrm{X}$$

There is a change in the relative motion of the projectile A and the target X but no internal excitation. Another possibility is that the projectile, the target or both can emerge in excited states denoted by A′, $X'$

$$\mathrm{A} + \mathrm{X} \rightarrow \mathrm{A}' + X'.$$

This is inelastic scattering. The third possibility is that the target and projectile are changed during the colision. For example, several nucleons (denoted by $a$) might be transferred from X to A to B and Y

$$\mathrm{A} + \mathrm{X} = \mathrm{A} + (\mathrm{Y} + a) \rightarrow (\mathrm{A} + a) + \mathrm{Y} = \mathrm{B} + \mathrm{Y}$$

This is a rearrangement collision or a reaction. In the following we denote the combined initial internal state of target and projectile by $|\alpha\rangle$. This state has internal energy $\varepsilon_\alpha$. In most applications the initial state is the ground state $|0\rangle$ and it is useful to choose $\varepsilon_0 = 0$. The final combined state of the target and projectile is denoted by $|\beta\rangle$ and has energy $\varepsilon_\beta$. If the reduced mass of the nuclei in the initial state is $\mu_\alpha$ and the relative momentum is $p_\alpha = \hbar k_\alpha$ then the kinetic energy of relative motion is $E_\alpha = \frac{1}{2}\hbar^2 k_\alpha^2/\mu_\alpha$. Similarly, in the final state $E_\beta = \frac{1}{2}\hbar^2 k_\beta^2/\mu_\beta$. Conservation of energy gives

$$E_\alpha + \varepsilon_\alpha = E_\beta + \varepsilon_\beta = E \tag{9.1}$$

where $E$ is the total energy.

The asymptotic form of the total scattering wave function in the channel $|\beta\rangle$ is

$$\left.\begin{aligned}\psi_\beta(\mathbf{r}) &\sim |\beta\rangle(1/r)f_{\beta\alpha}(\theta)\exp(ik_\beta r); \quad \text{if } \beta \neq \alpha \\ &\sim |\alpha\rangle[\exp(ik_\alpha z) + (1/r)f_{\alpha\alpha}(\theta)\exp(ik_\alpha r)]; \quad \text{if } \beta = \alpha\end{aligned}\right\} \tag{9.2}$$

In (9.2) $r$ is the relative coordinate of the centres-of-mass of the two nuclei and $f_{\beta\alpha}(\theta)$ is the scattering amplitude for scattering from state $|\alpha\rangle$ to $|\beta\rangle$ with centre-of-mass scattering angle $\theta$. The corresponding differential cross-section is

$$\sigma_{\beta\alpha}(\theta) = (v_\beta/v_\alpha)|f_{\beta\alpha}(\theta)|^2 \tag{9.3}$$

where $v_\alpha = \hbar k_\alpha/\mu_\alpha$ and $v_\beta = \hbar k_\beta/\mu_\beta$ are the relative velocities of the target and projectile in the initial and final channels.

We shall discuss inelastic scattering explicitly in the present chapter although many of the ideas can also be applied to reactions. The hamiltonian for the interacting system is

$$H = \mathbf{p}^2/2\mu + V_0(r) + h(\mathbf{r}, \xi) \tag{9.4}$$

$$h(\mathbf{r}, \xi) = H_0(\xi) + V(\mathbf{r}, \xi) \tag{9.5}$$

where $\mathbf{r}$ is the relative coordinate of the target and projectile and $\mathbf{p}$ is the conjugate momentum. The reduced mass $\mu$ is the same in the initial and final state. The internal coordinates of the target and projectile are denoted by $\xi$, $H_0(\xi)$ is the internal hamiltonian and $V(\mathbf{r}, \xi)$ is the coupling which produces the excitation. The potential $V_0(r)$ always contains the central part of the Coulomb interation between the target and the projectile; whereas the central part of the optical potential can be included either in $V_0(r)$ or in $V(\mathbf{r}, \xi)$. If the scattering problem with hamiltonian (9.4) is solved exactly the final results are identical. If approximate methods are used then different ways of dividing the interaction between $V_0(r)$ and $V(\mathbf{r}, \xi)$ yield different approximations.

## 9.2 Classical theory

Mott noticed already in 1931 that atomic collisions could be calculated in a simplified way if the relative motion of the two atoms was regarded as a classical degree of freedom. The excitation of internal quantal degrees of freedom is studied by solving time-dependent coupled equations. The Alder–Winther (1966) theory of multiple Coulomb excitation is based on the same idea. This theory describes the relative motion of target and projectile by a Rutherford orbit in the Coulomb potential $V_C(r)$. The excitation is obtained by solving the Schrödinger equation for the wave function of the internal coordinates $\xi$ evolving under the influence of a time-dependent interaction potential $V(\mathbf{r}(t), \xi)$ which is defined by the Rutherford orbit $\mathbf{r}(t)$ of relative motion. The amplitude $T_{\beta\alpha}\mathbf{r}(t)$ for a transition from the initial internal state $\alpha$ to the final internal state $\beta$ is written in terms of an evolution operator as

$$T_{\beta\alpha}[\mathbf{r}(t)] = \langle \beta | U[\mathbf{r}(t), t, t_0] | \alpha \rangle \tag{9.6}$$

where $U$ is a solution of the equation

$$i\hbar \partial U / \partial t = h(\mathbf{r}(t), \xi) U \tag{9.7}$$

with initial condition $U = 1$ when $t = t_0$. The probability of a transition from the initial state $\alpha$ to the final state $\beta$ for the path $\mathbf{r}(t)$ is $|T_{\beta\alpha}[\mathbf{r}(t)]|^2$ and the cross-section is calculated from the formula

$$\sigma_{\beta\alpha}(\theta) = \sigma_{\mathrm{R}}(\theta) | T_{\beta\alpha}[\mathbf{r}(\theta, t)]|^2 \tag{9.8}$$

where $\mathbf{r}(\theta, t)$ is the Rutherford orbit with scattering angle $\theta$.

In order for this classical approximation to be accurate it is ncessary that the Sommerfeld parameter $n$ be large. Also, the energy $\Delta E$ transferred from relative motion to internal excitation should be small compared with the total energy $E$ and the angular momentum transfer $\Delta\lambda$ should be small compared with the total angular momentum $\lambda$ of relative motion. Even if $\Delta E$ and $\Delta\lambda$ are small there is still some change in the orbit. Standard Coulomb excitation programmes solve (9.7) and calculate equation (9.8) using a symmetrized Rutherford orbit which is obtained by taking an average of the initial and final orbit parameters (de Boer *et al.*, 1977).

It is interesting to compare (9.8) with equation (4.5). If the optical potential is treated as a perturbation in elastic scattering then the elastic cross-section is

$$\sigma(\theta) = \sigma_{\mathrm{R}}(\theta) P_0[\mathbf{r}(\theta, t)] \tag{9.9}$$

where

$$P_0[\mathbf{r}(\theta, t)] = \exp\left\{ -(2/\hbar) \int_{-\alpha}^{\infty} W(\mathbf{r}(\theta, t))\, dt \right\} \tag{9.10}$$

is the probability that the interacting nuclei remain in their ground states after the scattering. Equation (9.9) is the same kind of approximation as (9.8) and the comparison suggests a way of obtaining the imaginary part of an optical potential which describes the depopulation of the ground state by inelastic scattering. We consider (9.8) for a ground state to ground state transition ($\alpha = 0$, $\beta = 0$) and obtain

$$\int_{-\alpha}^{\infty} W(\mathbf{r}(\theta, t)) dt = -(\hbar/2) \ln T_{00}[\mathbf{r}(\theta, t)] \tag{9.11}$$

This method has been used by Bonaccorso & Brink (1982) in a calculation of the imaginary part of the α-nucleus optical potential. Love, Terasawa & Satchler (1977) have shown that the effects of Coulomb excitation on elastic scattering can be simulated by a long range optical potential. This potential can be studied by using equation (9.11).

## 9.3 The Pechukas path integral

Although the approximate formula (9.8) for the inelastic cross-section is reasonable from a physical point of view and gives accurate results for multiple Coulomb excitation it has an unsatisfactory feature. As the theory treats the relative motion classically and the excitation by quantum mechanics there is no consistent way of improving it or of calculating corrections. In order to apply the theory in more general situations it is necessary to understand the sense in which (9.8) is an approximation to a full quantal calculation. This question can be studied by using a path integral approach to inelastic scattering which was developed by Pechukas (1969).

Pechukas writes a path integral for the amplitude $K_{\beta\alpha}(\mathbf{r}_1, t_1, \mathbf{p}_0 t_0)$ for the system to move from a relative momentum $\mathbf{p}_0$ at $t = t_0$ to a final position $\mathbf{r}_1$ at $t_1$ while the internal state changes from $\alpha$ to $\beta$. His path integral formula is

$$K_{\beta\alpha}(\mathbf{r}_1 t_1, \mathbf{p}_0 t_0) = \int D[\mathbf{r}(t)]\, T_{\beta\alpha}[\mathbf{r}(t)] \exp\{(i/\hbar) S_0[\mathbf{r}(t)]\} \tag{9.12}$$

The action $S_0[\mathbf{r}(t)]$ is given by equation (8.14)

$$S_0[\mathbf{r}(t)] = \int_{t_0}^{t_1} (\tfrac{1}{2}\mu \dot{\mathbf{r}}^2 - V_0(r))dt + \mathbf{p}_0 \cdot \mathbf{r}_0 \tag{9.13}$$

while $T_{\beta\alpha}[\mathbf{r}(t)]$ is the amplitude for a transition between the internal eigenstates $\alpha$ and $\beta$ as the colliding nuclei pass along $\mathbf{r}(t)$. This amplitude is calculated from (9.6) by solving (9.7) for a general path $\mathbf{r}(t)$ of relative motion. It is not a function of $t$ but depends on the specification of the entire path. The integral in (9.12) is over all paths satisfying the boundary conditions

$$\mu \dot{\mathbf{r}}(t_0) = \mathbf{p}_0, \quad \mathbf{r}(t_1) = \mathbf{r}_1 \tag{9.14}$$

The path integral (9.12) has been derived by Pechukas (1969) with different boundary conditions. He specifies the initial position rather than the initial momentum. Our choice is more suitable for use in scattering problems. We give a simple derivation of (9.12) using another method.

Consider a path integral for the propagator in all the variables

$$K(\mathbf{r}_1\xi_1 t_1, \mathbf{p}_0\xi_0 t_0) = \int\int D[\mathbf{r}(t)]D[\xi(t)] \times \exp\{(i/\hbar)(S_0[\mathbf{r}(t)] + S_1[\mathbf{r}(t), \xi(t)]\} \quad (9.15)$$

where $S_0$ is given by (9.13) and

$$S_1[\mathbf{r}(t), \xi(t)] = \int_{t_0}^{t_1} (p_\xi \dot{\xi} - h(\mathbf{r}(t), \xi(t))dt \quad (9.16)$$

We evaluate (9.15) by first integrating over the paths $\xi(t)$ of the internal coordinates to obtain

$$\langle \xi_1 | U[\mathbf{r}(t), t_1, t_0] | \xi_0 \rangle = \int_{t_0}^{t_1} D[\xi(t)] \exp\{(i/\hbar)S_1[\mathbf{r}(t), \xi(t)]\} \quad (9.17)$$

Equation (9.17) is a path integral for the operator $U$. But $U$ can also be calculated by conventional methods by solving the time-dependent Schrödinger equation (9.7). Substituting (9.17) in (9.15) gives

$$K(\mathbf{r}_1\xi_1 t_1, \mathbf{p}_0\xi_0 t_0) = \int D[\mathbf{r}(t)] \langle \xi_1 | U[\mathbf{r}(t), t_1, t_0] | \xi_0 \rangle \times \exp\{(i/\hbar)S_0[\mathbf{r}(t), \xi(t)]\} \quad (9.18)$$

Taking matrix elements of (9.18) between the internal states $|\alpha\rangle$ and $|\beta\rangle$ leads to the Pechukas integral (9.12).

The cross-section formula (9.8) used by Alder & Winther (1966) can be obtained by making the simplest stationary phase evaluation of the path integral (9.12). We take $V_0(r)$ in (9.13) to be the central part of the Coulomb interaction $V_C(r)$. Then we compare the path integral (9.12) with the multi-dimensional integral (B.15) in appendix B and identify $T_{\beta\alpha}$ with the slowly varying function $g(z)$ and $S_0$ with the phase $\psi(z)$. The stationary path is determined by the condition

$$\delta S_0[\mathbf{r}(t)] = 0$$

and is a Rutherford orbit $\mathbf{r}_C(t)$. The amplitude $T_{\beta\alpha}[\mathbf{r}_C(t)]$ is evaluated for that path and taken out of the integral. The resulting formula for the propagator is

$$K_{\beta\alpha}(\mathbf{r}_1 t_1, \mathbf{p}_0 t_0) = T_{\beta\alpha}[\mathbf{r}_C(t)] K_C(\mathbf{r}_1 t_1, \mathbf{p}_0 t_0) \quad (9.19)$$

where $K_C$ is the Coulomb propagator. The corresponding scattering wave function can be obtained from the propagator by generalizing the methods of section 8.2 and is

$$\psi_{\beta\alpha}(\mathbf{r}) \simeq t_{\beta\alpha}[\mathbf{r}_C(t)] \exp\{(i/\hbar)(\varepsilon_\alpha - \varepsilon_\beta)t_1\} \psi_C(\mathbf{r}) \quad (9.20)$$

where $\psi_C(\mathbf{r})$ is the elastic Coulomb wave function and $t_{\beta\alpha}$ is the transition amplitude calculated in the interaction representation

$$t_{\beta\alpha}[\mathbf{r}_C(t)] = T_{\beta\alpha}[\mathbf{r}_C(t)] \exp\{(i/\hbar)(\varepsilon_\beta t_1 - \varepsilon_\alpha t_0)\} \tag{9.21}$$

It is useful to choose the Coulomb orbit $\mathbf{r}_C(t)$ in (9.20) so that $t=0$ corresponds to the distance of closest approach. Then $t_{\beta\alpha}[\mathbf{r}_C(t)]$ in (9.20) is independent of the final position $\mathbf{r}$ and $t_1(\mathbf{r})$ is the time from the distance of closest approach to the final position.

The asymptotic form of the Coulomb wave function in (9.20) is

$$\psi_C(\mathbf{r}) \sim (1/r) f_C(E_\alpha, \theta) \exp\{i(k_\alpha r - n_\alpha \ln(2k_\alpha r))\}$$

while the inelastic wave function has the asymptotic form

$$\psi_{\beta\alpha}(\mathbf{r}) \sim (1/r) f_{\beta\alpha}(\theta) \exp\{i(k_\beta r - n_\beta \ln(2k_\beta r))\} \tag{9.22}$$

The phase of the scattered wave function (9.22) depends on the excitation energy $(\varepsilon_\beta - \varepsilon_\alpha)$ of the final state $|\beta\rangle$ through the final wave number $k_\beta$ and Sommerfeld parameter $n_\beta$. Dos Aidos & Brink (1984) show that the phase depending on $t_1$ in (9.20) gives the modification of the phase of the scattered wave function required by (9.22) to first order in the excitation energy $(\varepsilon_\beta - \varepsilon_\alpha)$. It also gives a modification of the phase of the Coulomb amplitude $f_C(\theta)$ which is equivalent to replacing the incident orbit energy $E_\alpha$ by an average $\frac{1}{2}(E_\beta + E_\alpha)$ of the initial and final energies. The final approximate form for the inelastic amplitude is

$$f_{\beta\alpha}(\theta) = f_C(\theta) t_{\beta\alpha}(\lambda_\theta) \tag{9.23}$$

Here $f_C(\theta)$ is the semi-classical Coulomb amplitude and $t_{\beta\alpha}(\lambda_\theta)$ is the transition amplitude (9.21) in the interaction representation calculated for an orbit with angular momentum $\lambda_\theta$ which is related to the scattering angle by the Rutherford formula (2.2). The final time $t_1$ has to be chosen so that $t=0$ corresponds to the point of closest approach in the Rutherford orbit. Some higher order effects can be included by calculating $f_C(\theta)$ and $t_{\beta\alpha}$ in (9.23) from an orbit with an average energy $\frac{1}{2}(E_\beta + E_\alpha)$.

Equation (9.23) is a good approximation to the inelastic scattering amplitude only if the excitation energy $\Delta E$ is small compared with the incident energy. In this limit the velocity ratio $v_\beta/v_\alpha \simeq 1$ in equation (9.3) and equation (9.23) leads to a cross-section formula (9.8) of the theory of Alder & Winther. Sukumar & Brink (1983) have developed a systematic method for calculating corrections to (9.23) and dos Aidos, Sukumar & Brink (1984) have evaluated the leading corrections for multiple Coulomb excitation. They show that some of the corrections can be incorporated by using a symmetrized orbit, but others must be calculated by different methods.

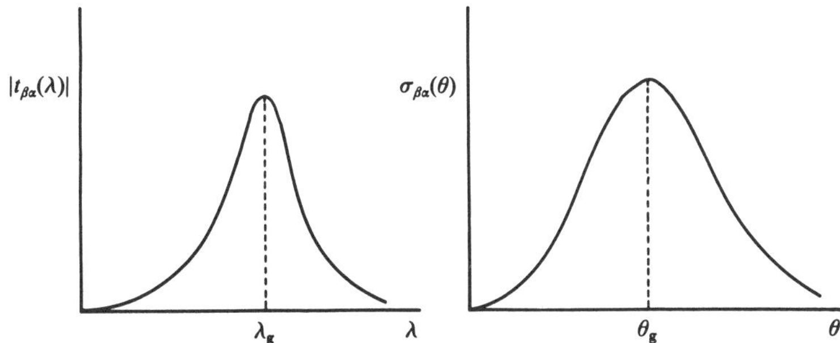


Fig. 9.1. (*a*) Sketch of the magnitude of the transition amplitude $|t_{\beta\alpha}(\lambda)|$ as a function of the angular momentum $\lambda$. (*b*) Sketch of the inelastic cross-section $\sigma_{\beta\alpha}(\theta)$ as a function of the scattering angle $\theta$.

The amplitude (9.23) and the corresponding cross-section formula are used widely for analysing the results of multiple Coulomb excitation experiments. In these applications the incident energy of the projectile is below the Coulomb barrier and the two nulcei do not come within the range of nuclear forces during the scattering. The interaction $V(\mathbf{r}, \xi)$ which produces the excitation is the non-central part of the Coulomb interaction due to deformations of the charge distributions of the target or projectile. Standard procedures have been developed for calculating the transition amplitude (see Alder & Winther, 1966), and information about the structure of the nuclei can be obtained by comparing calculated cross-sections with the experimental data.

The same method can be used at higher energies when nuclear reactions contribute to the excitation. If $V_0(r) = V_C(r)$ is the central part of Coulomb interaction then $V(\mathbf{r}, \xi)$ includes the central part of the optical potential and a nuclear part in the interaction causing excitation in addition to the non-central part of the Coulomb interaction. The transition amplitude $t_{\beta\alpha}$ factorizes

$$t_{\beta\alpha}(\lambda) = A_{\beta\alpha}(\lambda) \exp(2i\delta_n(\lambda)) \tag{9.24}$$

where $A_{\beta\alpha}(\lambda)$ is an excitation amplitude and $\delta_n(\lambda)$ is the semi-classical nuclear phase shift due to the optical potential calculated by the perturbation formula (3.63). The transition probability is

$$|t_{\beta\alpha}(\lambda)|^2 = |A_{\beta\alpha}(\lambda)|^2 P_0(\lambda) \tag{9.25}$$

where $P_0(\lambda)$ is the attenuation factor (9.10) due to the imaginary part of the optical potential. Fig. 9.1(*a*) shows a sketch of $|t_{\beta\alpha}(\lambda)|^2$ as a function of $\lambda$. It has a peak near the grazing angular momentum $\lambda_g$. The transition probability decreases for $\lambda > \lambda_g$ because the interaction becomes weak when the projectile passes far from the target and the excitation amplitude is small. When $\lambda < \lambda_g$ the probability $|t_{\beta\alpha}(\lambda)|^2$ decreases because of the attenuation factor $P_0(\lambda)$ which represents absorption by the optical potential. The excitation amplitudes are limited by unitarity

$$\sum_\beta |A_{\beta\alpha}(\lambda)|^2 \leqslant 1 \tag{9.26}$$

When $\lambda < \lambda_g$ the interaction is strong, the excitation is spread over many final states $\beta$ and each individual amplitude becomes small. This effect also contributes to the decrease in $|t_{\beta\alpha}(\lambda)|^2$ when $\lambda < \lambda_g$. The cross-section calculated from equation (9.8) is sketched in fig. 9.1(*b*). The grazing peak for $\theta \simeq \theta_g$ corresponds to the peak in the transition probability in fig. 9.1(*a*). The decrease in $\sigma_{\beta\alpha}(\theta)$ for $\theta > \theta_g$ is due to attenuation while for $\theta < \theta_g$ it is due to the weak interaction and small excitation amplitude.

A more accurate method for calculating inelastic scattering is to put

$$V_0(r) = V_C(r) + \text{Re}\, V_{\text{opt}}(r) \tag{9.27}$$

Then $t_{\beta\alpha}(\lambda)$ is arrived at by solving the time-dependent coupled equations (9.7) along a path $\mathbf{r}(t)$ with angular momentum $\lambda$ in the potential $V_0(r)$ and (9.23) is replaced by

$$f_{\beta\alpha}(\theta) = \sum_s f_s(\theta) t_{\beta\alpha}(\lambda_s) \tag{9.28}$$

In (9.28) $f_s(\theta)$ is a semi-classical elastic scattering amplitude in the potential $V(r)$. It can be calculated from equation (4.30). The sum over $s$ allows for the possibility that several classical orbits can have the same scattering angle. This possibility was discussed in detail in sections 4.4, 4.6 and 5.6. A particular example is a generalization of rainbow scattering which is often encountered with inelastic scattering experiments. Fig. 9.2 shows an example taken from Hansen *et al.* (1981). The interpretation is as follows: the elastic scattering of $^{14}C$ by $^{88}Sr$ shows a Coulomb rainbow pattern. When $\theta$ is less than the rainbow angle the inelastic amplitude is a sum of two parts

$$f_{\beta\alpha}(\theta) = f_1(\theta) t_{\beta\alpha}(\lambda_1) + f_2(\theta) t_{\beta\alpha}(\lambda_2) \tag{9.29}$$

The first term in (9.29) is due to the Coulomb excitation of the target. The angular momentum $\lambda_1$ is relatively large so that nuclear forces do not contribute. The second term comes from a grazing orbit which is deflected towards forward angles by the real part of the optical potential. Both Coulomb and nuclear forces contribute to the excitation. The characteristic structures in the inelastic cross-section forward of the grazing angle are due to interference between the two amplitudes in (9.29). Fig.

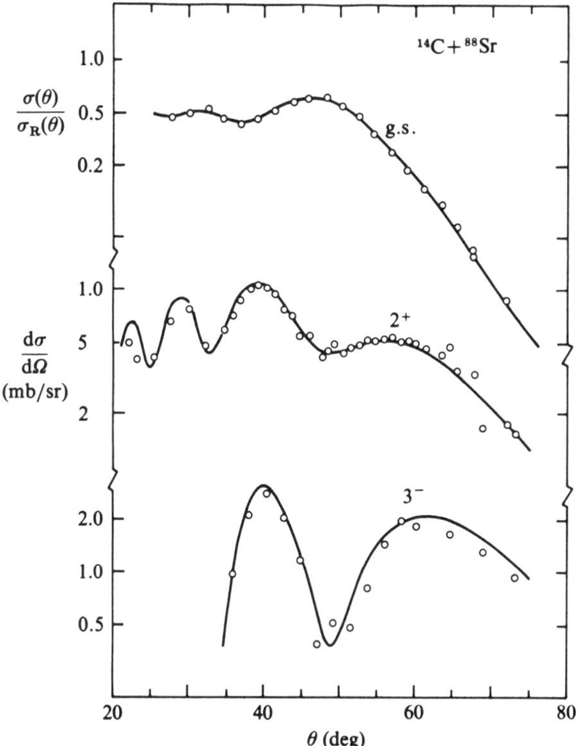


Fig. 9.2. Differential cross-sections for elastic and inelastic scattering of $^{14}C$ by $^{88}Sr$ at a laboratory energy of 51 MeV. The curves correspond to elastic scattering (g.s.), excitation of the $2^+$ state at 1.89 MeV and of the $3^-$ state at 2.73 MeV (from Hansen *et al.*, 1981).

9.2 is an example of a '180°' phase rule for Coulomb–nuclear interference phenomena. The rule states that oscillations observed for the excitation of different multipoles should be in phase with each other and out of phase with the oscillations in the elastic scattering. The phase rule holds because the excitation amplitudes in (9.29) have opposite signs. This is because $t_{\beta\alpha}(\lambda_1)$ is due to the Coulomb interactions and $t_{\beta\alpha}(\lambda_2)$ is due to the nuclear interactions; and the Coulomb and nuclear interactions have opposite signs. The phase rule was originally derived by Malfliet, Landowne & Rostokin (1973). It holds only if the excitation is a first-order process so that the amplitude is proportional to the interaction strength. Frahn (1976) has discussed the influence of absorption on the phase rule.

The decomposition (9.29) breaks down for scattering angles near to or greater than the rainbow angle. Then (9.29) has to be replaced by a more general uniform approximation or the diffraction methods of Frahn have to be used. We give a diffraction formula for the inelastic amplitude in section 9.4.

## 9.4 A diffraction formula for inelastic scattering

Equation (9.23) and its generalization (9.29) are stationary phase formulae for the inelastic scattering amplitude. Equation (9.29) includes interference effects but not diffraction and both (9.23) and (9.29) break down on the caustics at $\theta = 0$ and $\theta = \pi$. They also fail near a rainbow caustic. There are uniform approximations which remove the divergences at caustics and which are direct generalizations of the diffraction formulae (8.38) and (8.40) for elastic scattering. First we discuss the generalization of (8.38) which is valid for all except backward angles. Equation (8.38) contains a sum over $m$ which allows for the possibility that some trajectories orbit around the target several times. Normally only the $m = 0$ term is important and we omit the other terms in the discussion in this section. They can be added easily if required. The generalization of (8.38) is

$$f_{\beta\alpha}(\theta) = \frac{1}{(2\pi ik)}\left(\frac{\theta}{\sin\theta}\right)^{\frac{1}{2}} \int_0^\infty \int_0^{2\pi} \lambda d\lambda d\phi t_{\beta\alpha}(\lambda, \phi) e^{i(2\sigma(\lambda) - \lambda\theta\cos\phi)} \tag{9.30}$$

Equation (9.30) is written in a coordinate system with $z$-axis parallel to the incident beam direction $\mathbf{k}_\alpha$ and $y$-axis in the direction of $\mathbf{k}_\alpha \times \mathbf{k}_\beta$ where $\mathbf{k}_\beta$ is the wave vector of the scattered particle (fig. 9.3). These are standard axes for distorted wave Born approximation calculations and we refer to them as DWBA axes. The amplitude $t_{\beta\alpha}(\lambda, \phi)$ is calculated for a Rutherford orbit with angular momentum $\lambda$ and the plane of the orbit makes an angle $\phi$ with the $(x, z)$ plane. The time origin is chosen so that $t = 0$ at the distance of closest approach between the target and the projectile. Fig 9.3 shows a typical nearside orbit ($\phi = 0$) and a farside orbit ($\phi = \pi$). The integral (9.30) is non-singular on the caustic at $\theta = 0$. The stationary phase evaluation of the integral treats $t_{\beta\alpha}(\lambda, \phi)$ as a slowly varying function of $\lambda$ and $\phi$ and reduces to equation (9.23). Hence (9.30) satisfies the conditions for a uniform approximation. Semi-classical approximations to DWBA, which are special cases of (9.30), have been derived by other methods. We discuss this point in section 9.6.

Equation (9.30) can be written in various ways. The nuclear phase due to the optical potential can be separated from $t_{\beta\alpha}$ using (9.24), and (9.30)

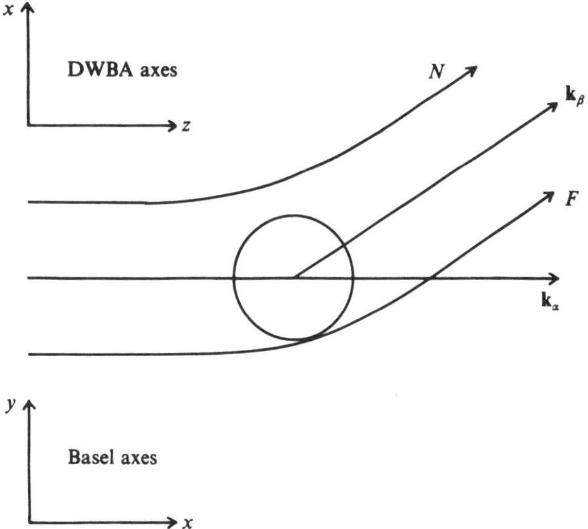


Fig. 9.3. Definition of DWBA axes and Basel axes. The diagram also shows a nearside ($N$) and a farside ($F$) orbit.

takes the form

$$f_{\beta\alpha}(\theta) = \frac{1}{2\pi ik}\left(\frac{\theta}{\sin\theta}\right)^{\frac{1}{2}} \int_0^\infty \int_0^{2\pi} \lambda d\lambda d\phi A_{\beta\alpha}(\lambda, \phi) e^{i(2\delta(\lambda) - \lambda\theta\cos\phi)} \tag{9.31}$$

Here $\delta(\lambda) = \sigma(\lambda) + \delta_n(\lambda)$ is the total phase shift. The nuclear phase $\delta_n(\lambda)$ is calculated by the perturbation formula (3.63). More accurate formulae can be obtained by using improved methods for calculating the excitation amplitude $A_{\beta\alpha}$ or the phase $\delta(\lambda)$. For example, the time-dependent coupled equations for $A_{\beta\alpha}$ can be solved for an orbit deflected by the real part of the optical potential. This allows for the effects of the attractive nuclear potential on the excitation. The nuclear phase $\delta_n(\lambda)$ could be improved by using the more accurate semi-classical formula (3.49) rather than the perturbation formula (3.63).

The energy of relative motion changes for an inelastic collision because some energy is transferred from relative motion into internal energy. The phase $\delta(\lambda)$ and the transfer amplitude $A_{\beta\alpha}(\lambda)$ should be calculated for some suitable average of the initial and final orbits. One possibility is to calculate them for an orbit with an energy of relative motion $\frac{1}{2}(E_\alpha + E_\beta)$ which is the average of the initial and final energies. Another possibility is to put $2\delta(\lambda) = \delta_\alpha(\lambda) + \delta_\beta(\lambda)$ where $\delta_\alpha(\lambda)$ is calculated with the energy $E_\alpha$ of the initial orbit and $\delta_\beta(\lambda)$ with the final orbit energy $E_\beta$. These various prescriptions are equivalent if the excitation energy $\Delta E = \varepsilon_\beta - \varepsilon_\alpha$ is small enough. There is no general agreement on the best procedure to use in other cases when $\Delta E$ is large and the reaction is badly matched.

When the initial and final states $|\alpha\rangle$ and $|\beta\rangle$ are angular momentum eigenstates with quantum numbers $(J_\alpha, M_\alpha)$ and $(J_\beta, M_\beta)$ then the $\phi$ dependence of $A_{\beta\alpha}(\lambda, \phi)$ is given by

$$A_{\beta\alpha}(\lambda, \phi) = A_{\beta\alpha}(\lambda) e^{i(M_\alpha - M_\beta)\phi} \tag{9.32}$$

This is because the total $z$-component of angular momentum is conserved and the change in the $z$-component of orbital angular momentum is $M_\alpha - M_\beta$. When (9.32) is substituted into equation (9.31) the $\phi$-integral can be evaluated to give a Bessel function (appendix A.10) and

$$f_{\beta\alpha}(\theta) = \frac{i^{M-1}}{k}\left(\frac{\theta}{\sin\theta}\right)^{\frac{1}{2}} \int_0^\infty \lambda d\lambda A_{\beta\alpha}(\lambda) e^{2i\delta(\lambda)} J_M(\lambda\theta) \tag{9.33}$$

where $M = M_\beta - M_\alpha$. Formulae which are equivalent to (9.33) have been obtained by Potgieter & Frahn (1967), Dar (1966), Hahne (1967), Frahn (1980*a, b*) and others by making an approximate evaluation of the DWBA formula for an inelastic amplitude. It is often useful to write (9.31) and (9.34) in terms of the angular momentum transfer $(L, M)$. This is done by putting

$$f_{\beta\alpha}(\theta) = \sum_L f_{LM}^{\beta\alpha}(\theta) \langle J_\beta M_\beta | LMJ_\alpha M_\alpha \rangle \tag{9.34}$$

$$A_{\beta\alpha}(\lambda) = \sum_L A_{LM}(\lambda) \langle J_\beta M_\beta | LMJ_\alpha M_\alpha \rangle \tag{9.35}$$

where the factors written with angular brackets are Clebsch–Gordan coefficients. Then (9.31) becomes

$$f_{LM}(\theta) = \frac{1}{(2\pi ik)}\left(\frac{\theta}{\sin\theta}\right)^{\frac{1}{2}} \int_0^\infty \int_0^{2\pi} \lambda d\lambda d\phi A_{AM}(\lambda) \cdot \exp(-iM\phi) \exp(i(2\delta(\lambda) - \lambda\theta\cos\phi)) \tag{9.36}$$

There is an analogous formula corresponding to equation (9.33). The total cross-section summed over the final angular momentum projection and averaged over the initial angular momentum projection is

$$\begin{aligned}\sigma(\theta) &= \frac{1}{(2J_\alpha + 1)} \frac{v_\beta}{v_\alpha} \sum_{M_\alpha M_\beta} |f_{\beta\alpha}(\theta)|^2 \\ &= \frac{(2J_\beta + 1)}{(2J_\alpha + 1)} \frac{v_\beta}{v_\alpha} \sum_{LM} \frac{1}{(2L+1)} |f_{LM}(\theta)|^2\end{aligned} \tag{9.37}$$

The amplitudes $f_{LM}$ satisfy a symmetry condition which is a consequence of reflection symmetry in the reaction plane (Bohr, 1959). It is

$$f^{\beta\alpha}_{L-M}(\theta) = \pi_\alpha \pi_\beta (-1)^{L-M} f^{\beta\alpha}_{LM}(\theta) \tag{9.38}$$

where $\pi_\alpha$ and $\pi_\beta$ are the parities of the states $\alpha$ and $\beta$.

When the scattering angle $\theta$ is large the integral over $\phi$ in (9.36) can be calculated by the stationary phase approximation. There are two stationary points at $\phi = 0$ and $\phi = \pi$ and using the stationary phase formula (B.9) we get

$$f_{LM}(\theta) = f^{-}_{LM}(\theta) + (-1)^M f^{+}_{LM}(\theta) \tag{9.39}$$

$$f^{\pm}_{LM}(\theta) = \frac{1}{ik}\int_0^\infty \left(\frac{\lambda}{2\pi \sin\theta}\right)^{\frac{1}{2}} d\lambda A_{LM}(\lambda) e^{i(2\delta(\lambda) \pm \lambda\theta \mp \frac{1}{4}\pi)} \tag{9.40}$$

The nearside amplitude $f^{-}_{LM}(\theta)$ comes from the stationary point $\phi = 0$, while the farside amplitude is the contribution from $\phi = \pi$. The factor $(-1)^M$ comes from the term $e^{-iM\phi}$ in (9.36) evaluated at $\phi = \pi$. The labels $\alpha$ and $\beta$ for the initial and final states have been omitted in (9.39) and (9.40) to simplify the notation. Equation (9.40) is very similar to the $m = 0$ term in the Poisson sum formula (4.17) for the near- and farside components of the elastic scattering amplitude.

Often it is convenient to write (9.39) and (9.40) with respect to Basel axes (fig. 9.3). The $x$-axis is parallel to the incident beam direction and the $z$-axis in the direction $\mathbf{k}_\alpha \times \mathbf{k}_\beta$ perpendicular to the reaction plane. The scattering amplitude $\mathscr{F}_{LM}$ in the new coordinate system is related to the $f_{LM}$ by a rotation

$$\mathscr{F}_{LM}(\theta) = (-1)^M \sum_K i^K d^L_{KM}(\tfrac{1}{2}\pi) f_{LM}(\theta) \tag{9.41}$$

(see Frahn, 1980*b*). The $d^L_{KM}(\frac{1}{2}\pi)$ are rotation matrices (Brink & Satchler, 1968; Messiah, 1959). In the new axes (9.39) and (9.40) take the form

$$\mathscr{F}_{LM}(\theta) = \mathscr{F}^{-}_{LM}(\theta) + (-1)^L \mathscr{F}^{+}_{LM}(\theta) \tag{9.42}$$

$$\mathscr{F}^{\pm}_{LM}(\theta) = \frac{1}{ik}\int_0^\infty \left(\frac{\lambda}{2\pi \sin\theta}\right)^{\frac{1}{2}} d\lambda B_{L\mp M}(\lambda) e^{i(2\delta(\lambda) \pm \lambda\theta \mp \frac{1}{4}\pi)} \tag{9.43}$$

The transition amplitudes $B_{LM}(\lambda)$ in the Basel axes are obtained from the $A_{LM}(\lambda)$ by a transformation like (9.41). The structure of the farside term in (9.42) can be obtained by making the transformation (9.41) explicitly and using symmetry properties of the rotation matrices. It can also be derived by noting that the farside contribution is obtained by a rotation through

180° about the $x$-axis in the Basel coordinate system. For such a rotation

$$B_{LM}(\lambda)\to(-1)^{L}B_{L-M}(\lambda) \tag{9.44}$$

A formula equivalent to (9.42) and (9.43) has been used recently by Dean & Rowley (1984) for a semi-classical analysis of inelastic heavy-ion scattering. Bohr's symmetry property (9.38) implies that

$$\mathscr{F}_{LM}(\theta)=(-1)^{M}\pi_{\alpha}\pi_{\beta}\mathscr{F}_{LM}(\theta) \tag{9.45}$$

so that $\mathscr{F}_{LM}(\theta)$ is zero unless $(-1)^{M}=\pi_{\alpha}\pi_{\beta}$.

We conclude this section by giving the diffraction formula for $f_{LM}(\theta)$ which is analogous to the elastic scattering formula (8.40) and which is valid for back-angle scattering. We write it in terms of $\vartheta=\pi-\theta$

$$f^{\beta\alpha}_{LM}(\theta)=f^{\beta\alpha}_{LM}(\vartheta)$$
$$=\frac{1}{2\pi ik}\left(\frac{\vartheta}{\sin\vartheta}\right)^{\frac{1}{2}}\sum_{n=-\infty}^{\infty}(-1)^{n}\int\int\lambda d\lambda d\phi A_{LM}(\lambda)e^{-iM\phi}e^{i\bar{F}_{n}(\lambda,\phi,\vartheta)} \tag{9.46}$$

where $\bar{F}_{n}$ is defined in equation (8.41). Equation (9.46) is a generalization of formula (5.45) for the elastic amplitude. In section 5.7 we found that the $n=0$ term in equation (5.45) was important for scattering near the Coulomb barrier while the $n=1$ term contributed at higher energies. These same features hold for back-angle inelastic scattering calculated from (9.46).

## 9.5 Blair's phase rule

In 1959 Blair developed a simple theory of inelastic diffractive scattering of $\alpha$-particles by nuclei. He considered the excitation of collective states characterized by deformation parameters $a_{LM}$. The scattering amplitude $f_{LM}(\theta)$ for a transition from a ground state $|0\rangle$ with zero spin to a one-phonon state $|LM\rangle$ with angular momentum quantum numbers $(L, M)$ was calculated using the adiabatic approximation. In the adiabatic limit the excitation energy $\Delta E$ of the state $|L, M\rangle$ is negligible and the corresponding oscillation frequency $\Delta E/\hbar$ is so small that the deformation coordinate $a_{LM}$ does not change during the collision. The scattering amplitude in the adiabatic approximation is

$$f_{LM}(\theta)=\langle LM|f(\theta,a)|0\rangle \tag{9.47}$$

where $f(\theta,a)$ is the elastic amplitude for fixed deformation $a_{LM}$. Blair considered the nucleus to be strongly absorbing with a sharp radius

$$R(\theta,\phi)=R_{0}\left(1+\sum_{LM}a_{LM}Y^{*}_{LM}(\theta,\phi)\right) \tag{9.48}$$

and he calculated the scattering by a Fraunhofer approximation neglecting

Coulomb effects. Blair's model provides a very simple example of the theory derived in section 9.4.

The elastic amplitude when $a_{LM} = 0$ is the familiar result for scattering by a black disk (2.29)

$$f(\theta) = (i\lambda_0/k\theta)J_1(\lambda_0\theta) \tag{9.49}$$

where $\lambda_0 = kR_0$. The inelastic amplitude can be obtained from equation (9.33). The excitation amplitude $A_{LM}(\lambda)$ is peaked at the grazing angular momentum $\lambda_0$ and can be approximated by a delta function

$$A_{LM}(\lambda) = -A_L Y^*_{LM}(\tfrac{1}{2}\pi, 0)\lambda_0\delta(\lambda - \lambda_0) \tag{9.50}$$

where $A_L$ is a reduced matrix element of $a_{LM}$. The diffraction is determined by the shape of the nuclear surface $R(\frac{1}{2}\pi, \phi)$ on the plane perpendicular to the incident beam. This is given by (9.48) with $\theta = \frac{1}{2}\pi$ and is the origin of the factor $Y^*_{LM}(\frac{1}{2}\pi, 0)$ in (9.50). Substituting into (9.33) and neglecting the phase $\delta(\lambda)$ gives

$$f_{LM}(\theta) = i^{M+1}(\lambda_0 A_L/k)J_M(\lambda_0\theta)Y^*_{LM}(\tfrac{1}{2}\pi, 0) \tag{9.51}$$

We have assumed that the scattering angle is small so that $\theta/\sin\theta \simeq 1$. The cross-section corresponding to (9.51) is

$$\sigma_L = R_0^2 A_L^2 \sum_M |Y_{LM}(\tfrac{1}{2}\pi, 0)|^2 J_M^2(\lambda_0\theta) \tag{9.52}$$

Both the elastic cross-section calculated from (9.49) and the inelastic cross-section (9.52) oscillate with angle. The asymptotic form (A.5) for Bessel functions for large $\lambda_0\theta$ gives

$$J_M^2(\lambda_0\theta) \sim (2/\pi\lambda_0\theta)\sin^2(\lambda_0\theta + \tfrac{1}{4}\pi); \quad M \text{ even} \tag{9.53}$$

$$\sim (2/\pi\lambda_0\theta)\cos^2(\lambda_0\theta + \tfrac{1}{4}\pi); \quad M \text{ odd} \tag{9.54}$$

Thus the phase of the oscillations in $J_M^2(\lambda_0\theta)$ depends on whether $M$ is even or odd. The spherical harmonics $Y_{LM}(\frac{1}{2}\pi, 0) = 0$ if $L + M$ is odd. Hence the sum (9.52) contains only odd values of $M$ if $L$ is odd. The cross-section $\sigma(\theta)$ has an angle dependence given by (9.54) and is in phase with the elastic cross-section proportional to $J_1^2(\lambda_0\theta)$. When $L$ is even the sum (9.52) contains only even values of $M$ and the oscillations in $\sigma_L(\theta)$ are out of phase with those in the elastic cross-section. This is the Blair phase rule.

It is also interesting to study the Blair phase rule and to use the amplitudes (9.42) and (9.43) in the Basel axes (Dean & Rowley, 1984). The transition amplitudes $B_{LM}$ are given by

$$\begin{aligned} B_{LM}(\lambda) &= -A_L\lambda_0 Y^*_{LM}(\tfrac{1}{2}\pi, \tfrac{1}{2}\pi)\delta(\lambda - \lambda_0) \\ &= -A_L\lambda_0 i^{-M} Y^*_{LM}(\tfrac{1}{2}\pi, 0)\delta(\lambda - \lambda_0) \end{aligned} \tag{9.55}$$

The symmetry properties of spherical harmonics can be used to show that

$$B_{L-M}(\lambda) = B_{LM}(\lambda) \tag{9.56}$$

The cross-section is given by

$$\sigma_L(\theta) = \sum_M |\mathscr{F}^-_{LM}(\theta) + (-1)^L \mathscr{F}^+_{LM}(\theta)|^2 \tag{9.57}$$

Equation (9.57) shows that the oscillations in the cross-section are due to the interference between the nearside amplitude $\mathscr{F}^-$ and the farside amplitude $\mathscr{F}^+$. The symmetry relation (9.56) which holds for the Blair model shows that $\mathscr{F}^+_{LM} = \mathscr{F}^+_{L-M}$ and

$$\sigma_L(\theta) = \left(\frac{\theta_0}{2\pi k^2 \sin\theta}\right) \sum_M |Y_{LM}(\tfrac{1}{2}\pi, 0)|^2 |e^{-i(\lambda_0\theta - \frac{1}{4}\pi)} + (-1)^L e^{i(\lambda_0\theta - \frac{1}{4}\pi)}|^2 \tag{9.58}$$

This result is equivalent to (9.52) when $\lambda_0\theta$ is large and shows that, in the Basel axes, the Blair phase rule is a consequence of the factor $(-1)^L$ in equation (9.57).

## 9.6 Applications

We now give some examples of applications of the diffraction formulae in section 9.4. We begin the discussion with equation (9.42) in the Basel axes. Blair's model (section 9.5) neglects Coulomb effects and makes some very special assumptions about the transfer amplitude $B_{LM}(\lambda)$: it is proportional to a delta function in $\lambda$ peaked at the grazing angular momentum, its dependence on $M$ is proportional to a spherical harmonic and it has the special symmetry (9.56). A more realistic amplitude which is general enough to cover many examples of inelastic scattering and reactions is obtained by writing

$$B_{LM}(\lambda)\exp(2i\delta_n(\lambda)) = B_{LM}\, g(\lambda - \lambda_g) \tag{9.59}$$

where $g(\lambda - \lambda_g)$ is peaked around the grazing angular momentum $\lambda_g$. The main assumption in (9.59) is that the $\lambda$ dependence is the same for all magnetic substates $M$. Bohr's symmetry property (9.38) implies that $B_{LM} = 0$ unless $(-1)^M = \pi_\alpha \pi_\beta$ where $\pi_\alpha$ and $\pi_\beta$ are the parities of the initial and final states. We obtain expressions for the inelastic amplitude by substituting (9.59) into the diffraction formula (9.43). Following Strutinsky (1964) we approximate the integral by putting $\sqrt{\lambda} = \sqrt{\lambda_g}$ and expanding the Coulomb phase about $\lambda_g$ to first order

$$2\sigma(\lambda) \simeq 2\sigma(\lambda_g) + (\lambda - \lambda_g)\theta_g \tag{9.60}$$

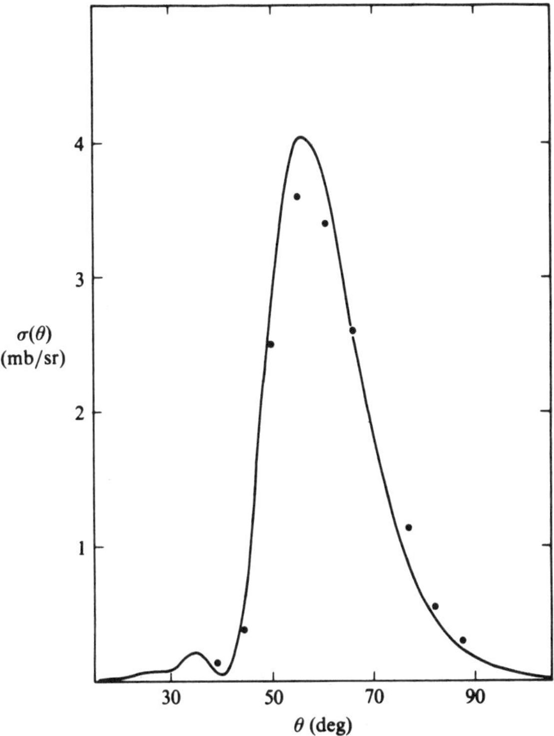


Fig. 9.4. Angular distribution for the $^{94}$Mo ($^{13}$C, $^{12}$C) $^{95}$Mo single-neutron transfer reaction at a laboratory energy of 51 MeV. The final nucleus is formed in its ground state (from Bond *et al.*, 1974).

where $\theta_g$ is the Coulomb grazing angle and is related to $\lambda_g$ by the Rutherford formula (2.2). Then the nearside and farside components of the amplitude (9.42) can be written in the form

$$\mathscr{F}^{\pm}_{LM}(\theta) \simeq B_{L\mp M} C^{\pm}(\theta)\tilde{g}(\theta_g \pm \theta) \tag{9.61}$$

$$C^{\pm}(\theta) = \frac{1}{ik}\left(\frac{\lambda_g}{2\pi \sin\theta}\right)^{\frac{1}{2}} e^{i(2\sigma(\lambda_g) \pm \theta\lambda_g \mp \frac{1}{4}\pi)} \tag{9.62}$$

where $\tilde{g}$ is the Fourier transform of the form factor $g$

$$\tilde{g}(\chi) = \int_{-\infty}^{\infty} e^{i\chi\mu} g(\mu) d\mu \tag{9.63}$$

Strutinsky pointed out that the nearside amplitude gives the dominant contribution to the cross-section for reactions with a large Sommerfeld parameter so that

$$\sigma_L(\theta) \simeq \sum_M |B_{LM}|^2 |C^-(\theta)\tilde{g}(\theta_g - \theta)|^2 \tag{9.64}$$

This cross-section is peaked near the grazing angle $\theta_g$. An example of this kind of cross-section is shown in fig. 9.4.

The width $\Delta\theta$ of the grazing peak has been discussed by Siemens & Becchetti (1972) and Strutinsky (1973). These authors show how $\Delta\theta$ depends on the width $\Delta$ of the form factor $g(\lambda - \lambda_g)$ in angular momentum space. Strutinsky uses a simple gaussian form factor

$$g(\lambda - \lambda_g) = \exp\{-(\lambda - \lambda_g)^2/\Delta^2\}$$

This form factor has the advantage that the Fourier transform can still be calculated even when one more term is retained in the expansion (9.60) of the Coulomb phase (see equation 5.12). Then the square of the Fourier transform in equation (9.64) can be written as

$$|\tilde{g}(\theta_g - \theta)|^2 = \frac{\pi\Delta^2}{1 + \kappa^2\Delta^4} \exp\left\{-\frac{1}{2}\frac{\Delta^2(\theta_g - \theta)^2}{1 + \kappa^2\Delta^4}\right\} \tag{9.65}$$

where

$$\kappa = (1/n)\sin^2(\tfrac{1}{2}\theta_g)$$

In the limit, when $\Delta$ is small ($\kappa\Delta^2 \ll 1$), the grazing peak in the angular distribution is spread by diffraction and has a width inversely proportional to $\Delta$. In the classical limit, when $\Delta$ is large ($\kappa\Delta^2 \gg 1$), equation (9.65) shows that the width of the grazing peak is proportional to $\Delta$. In this limit the classical formula (9.8) for the cross-section can be used. When $\kappa\Delta^2 = 1$ the width of the grazing peak is a minimum. A simple estimate from equation (9.65) shows that the full width at half maximum is

$$(\Delta\theta)_{\min} \simeq (3/\sqrt{n})\sin(\tfrac{1}{2}\theta_g)$$

When $n \lesssim 1$ this formula shows that the grazing peak in the angular distribution is very broad ($\Delta\theta > \theta_g$). Then diffraction effects dominate and the classical character of the angular distribution is lost.

Even when the Sommerfeld parameter is quite large there are cases where both nearside and fraside amplitudes contribute to the cross-section and

$$\sigma_L(\theta) = \sum_M |\mathscr{F}^-_{LM}(\theta) + (-1)^L \mathscr{F}^+_{LM}|^2 \tag{9.66}$$

where $\mathscr{F}^{\pm}$ are given by equation (9.60). Fig. 9.5 shows an example for the reaction $^{18}O + {}^{60}Ni \to {}^{16}O + {}^{62}Ni$ at a laboratory energy of 65 MeV. In this example $n = 18$ and the grazing angular momentum $\lambda_g = 50$. The angular distribution shows a grazing peak with nearside-farside inter-

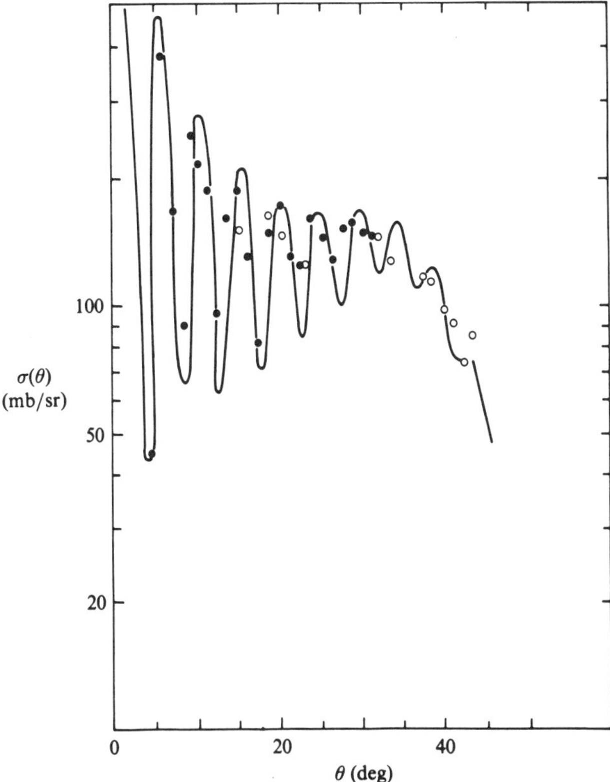


Fig. 9.5. Angular distribution for the two-neutron transfer reaction $^{60}Ni$ ($^{18}O$, $^{16}O$) $^{62}Ni$ to the ground state of the final nucleus at a laboratory energy of 65 MeV. The dots and circles show experimental data (from LeVine *et al.*, 1974).

ference oscillations for $\theta < \theta_g$. Kahana & Baltz (1977) have studied the transition from the classical to the diffractive angular distribution as $n$ decreases.

All the nuclear states involved in the example shown in fig. 9.5 have zero angular momenta so the transferred angular momentum is $L = 0$. This is the reason why the nearside–farside interference oscillations are so prominent. When $L > 0$ the interference oscillations are usually damped. This is because the spins of the final nuclei for a transfer reaction or for inelastic scattering tend to be polarized perpendicularly to the reaction plane. In order to discuss this effect we write

$$B_{LM} = i^M d^L_{0M}(\tfrac{1}{2}\pi)\rho(M) \tag{9.67}$$

The notation here is chosen to facilitate comparison with the work of

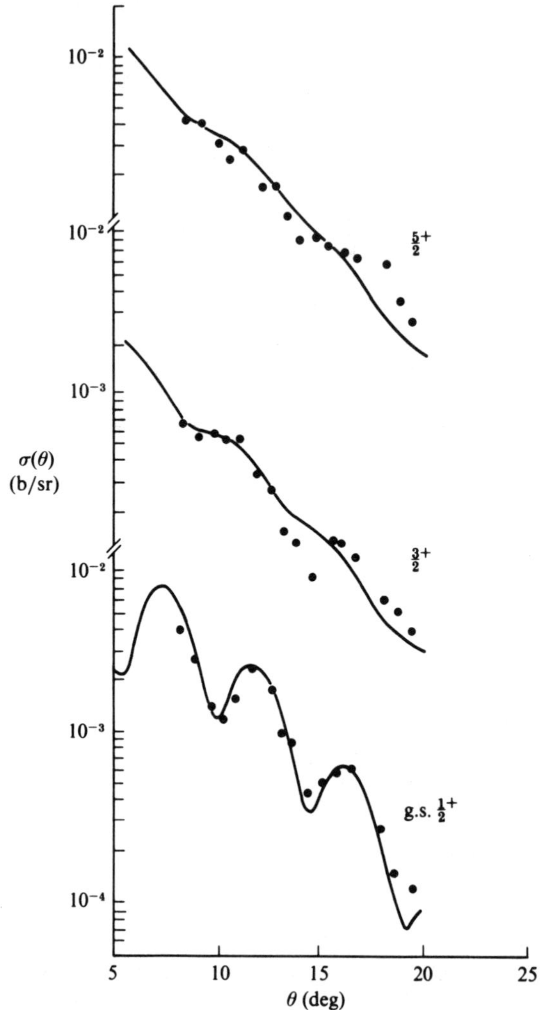


Fig. 9.6. Angular distributions for the neutron transfer reaction $^{26}$Mg ($^{11}$B, $^{10}$B) $^{27}$Mg at a laboratory energy of 114 MeV. The final states in $^{27}$Mg are the ground state ($\frac{1}{2}^+$) and excited states at 0.98 MeV ($\frac{3}{2}^+$) and 1.70 MeV ($\frac{5}{2}^+$). (The diagram is taken from Hasan & Brink, 1978; the data is from Paschopoulos, 1975.)

Hahne (1967), Frahn (1980) and other authors who use Frahn's diffraction methods. The rotation matrix in (9.67) is proportional to a spherical harmonic $Y_{LM}(\frac{1}{2}\pi, 0)$ and if the coefficients $\rho(M)$ are independent of $M$ then the amplitude (9.67) is equivalent to Blair's amplitude (9.56) in its dependence on $L$ and $M$. The amplitude (9.67) also satisfies Bohr's symmetry conditions (9.45) for a natural parity transition $(-1)^L = \pi_\alpha \pi_\beta$. If $\rho(M)$ is independent of $M$ then the discussion in section 9.5 shows that there are strong nearside–farside interference effects in the cross-section which satisfy the Blair phase rule. In order to understand the effects of polarization we notice that in (9.66) a nearside amplitude $B_{LM}$ interferes with a farside amplitude $B_{L-M}$. In an extreme case of strong polarization in the direction of the $z$-axis $B_{LM} = 0$ for $M < 0$. Then all the interference terms in (9.66) are zero and the cross-section is a smooth function of angle. This is an extreme case but it is clear that any polarization of the final state perpendicular to the reaction plane will tend to damp out the nearside–farside interference effects.

Fig 9.6 shows an example taken from Hasan & Brink (1978). The data is from Paschopoulos *et al.* (1975) and shows angular distributions for the neutron transfer reaction $^{26}$Mg ($^{11}$B, $^{10}$B) $^{27}$Mg at a laboratory energy of 114 MeV. Three final states are observed: the ground state with spin and parity $J^\pi = \frac{1}{2}$ (single particle shell model state $S_{\frac{1}{2}}$), at 0.98 MeV ($J^\pi = \frac{3}{2}$, $d_{\frac{3}{2}}$) and a state at 1.70 MeV($J^\pi = \frac{5}{2}$, $d_{\frac{5}{2}}$). The transferred nucleon comes from a $p_{\frac{3}{2}}$ state in $^{11}$B. The angular momentum transfer to the ground state of the final nucleus is $L = 1$ (equation 9.34) while $L = 3$ is possible for the two excited states. None of the three angular distributions shows any trace of a grazing peak. This is because the Sommerfeld parameter is small ($n = 2.9$). Neutron transfer tends to produce a polarization of the spin of the final state. For the $L = 1$ transfer to the ground state of $^{24}$Mg nearside–farside oscillations are observed, but they have almost disappeared in the transfer to the excited states where the spins are large and a larger polarization is possible.

The semi-classical theory of cross-sections like the one shown in fig. 9.6 has been discussed by Potgieter & Frahn (1967), Dar (1966), Hahne (1967), Hasan & Brink (1978, 1979), Frahn (1980*b*), Dean & Rowley (1984) and others. The starting point for these studies is usually the DWBA amplitude (see Frahn, 1980*b*)

$$f_{LM}(\theta) = \sum_{ll'} i^{-L} b_{l'l}^{LM} (2l+1)^{\frac{1}{2}} R_{ll'}^{L} \exp[i(\sigma_{l'} + \sigma_l)] Y_{l-M}(\theta, 0) \qquad (9.68)$$

$$b_{l'l}^{LM} = i^{l'-l} \langle lL00|l'0\rangle \langle lL-MM|l'0\rangle \qquad (9.69)$$

Here, the $R_{ll'}^{L}$ are radial integrals and $\sigma_{l'}$ and $\sigma_l$ are Coulomb phases for the entrance and exit channels. A diffraction formula is obtained from equation (9.68) by making several approximations. The angular momenta $l'$ and $l$ in the Clebsch–Gordan coefficients in (9.69) are assumed to be large and they are approximated by rotation matrices (Frahn, 1980*b*). Then (9.69) simplifies to

$$b_{l'l}^{LM} \simeq b_{KM}^{L} = i^K d_{0K}^L(\tfrac{1}{2}\pi) d_{MK}^L(\tfrac{1}{2}\pi) \tag{9.70}$$

where $K = l' - l$. The spherical harmonic is approximated by a Bessel function (Austern & Blair, 1965)

$$Y_{l-M}(\theta, 0) \simeq \left[\frac{\lambda}{2\pi}\frac{\theta}{\sin\theta}\right]^{\frac{1}{2}} J_M(\lambda\theta);\ 0 \leqslant \theta < \pi - 1/l \tag{9.71}$$

where $\lambda = l + \frac{1}{2}$. Finally the sum over $l$ in (9.68) is replaced by an integral over $\lambda$ using the Poisson sum formula. If we keep only the leading term the result is a diffraction formula like (9.33) (Frahn, 1980*b*)

$$f_{LM}(\theta) = \frac{i^{-L}}{\sqrt{\pi}}\left(\frac{\theta}{\sin\theta}\right)^{\frac{1}{2}} \sum_K b_{KM}^L \int_0^\infty \lambda d\lambda R_{LK}(\lambda) \times \exp\left[i(\sigma(\lambda) + \sigma(\lambda + K)\right] J_M(\lambda\theta) \tag{9.72}$$

where $R_{ll+K}^L = R_{LK}(\lambda)$. Equation (9.72) is equivalent to (9.33) if

$$A_{LM}(\lambda)\exp(2i\delta_n(\lambda)) = (k/\sqrt{\pi}) i^{1-L-M} \sum_K b_{KM}^L R_{LK}(\lambda) \exp(\tfrac{1}{2}iK\Theta(\lambda))$$

where we have used the approximation

$$\sigma(\lambda + K) \simeq \sigma(\lambda) + \tfrac{1}{2}K\Theta_C(\lambda)$$

and $\Theta_C(\lambda)$ is the Coulomb scattering angle for an orbit with angular momentum $\lambda$.

# 10
# Non-local potentials

## 10.1 Examples in heavy-ion scattering

In this section we give two examples of non-local potentials that arise in nuclear scattering problems. A general non-local potential $v$ is represented by an integral operator

$$v\psi \rightarrow \int v(\mathbf{r}, \mathbf{r}')\psi(\mathbf{r}')d\mathbf{r}'$$

where $v(\mathbf{r}, \mathbf{r}') = \langle \mathbf{r}|v|\mathbf{r}'\rangle$ is the matrix element of $v$ in the coordinate representation. A characteristic property is the range $\beta$ of non-locality. This is defined so that $v(\mathbf{r}, \mathbf{r}')$ becomes small if $|\mathbf{r} - \mathbf{r}'| \gtrsim \beta$. A local potential is a limiting case when $\beta = 0$ and

$$v(\mathbf{r}, \mathbf{r}') = v_0\delta(\mathbf{r} - \mathbf{r}')$$

In this chapter we consider a Schrödinger equation

$$[-(\hbar^2/2\mu)\nabla^2 + V_0(r)]\psi(\mathbf{r}) + \int v(\mathbf{r}, \mathbf{r}')\psi(\mathbf{r}')d\mathbf{r} = E\psi(\mathbf{r}) \tag{10.1}$$

The total potential has a local part $V_0(r)$ and a non-local part $v(\mathbf{r}, \mathbf{r}')$.

Feshbach's optical potential is an example of a non-local potential used in nuclear scattering problems. Feshbach (1958) starts with a coupled-channels hamiltonian of the form used in chapter 9

$$H = p^2/2\mu + V_0(\mathbf{r}) + H_0(\xi) + V(\mathbf{r}, \xi) \tag{10.2}$$

The relative coordinate $\mathbf{r}$ is coupled to internal degrees of freedom $\xi$ by an interaction $V(\mathbf{r}, \xi)$. Let $|\alpha\rangle$ denote an eigenstate of the intrinsic hamiltonian $H_0(\xi)$ with energy $\varepsilon_0$. Feshbach projects the complete hamiltonian on to the elastic channel and obtains an effective interaction for elastic scattering which contains implicitly all the effects due to excitation of intrinsic states during the scattering. His potential calculated to second order is

$$v(\mathbf{r}, \mathbf{r}') = \sum_{\alpha \neq 0} \langle 0|V(\mathbf{r}, \xi)|\alpha\rangle\langle\alpha|V(\mathbf{r}', \xi)|0\rangle G(\mathbf{r}, \mathbf{r}', E - \varepsilon_\alpha) \tag{10.3}$$

where $G(\mathbf{r}, \mathbf{r}', E)$ is the propagator (7.31) for the relative motion with energy

$E$ in the zero order potential $V_0(r)$. The physical meaning of (10.3) is the following: The system starts in the ground state at $\mathbf{r}'$. Then it is excited to the state $|\alpha\rangle$ by $V$. It propagates in that state to $\mathbf{r}$ and is de-excited to the ground state. The potential $v(\mathbf{r}, \mathbf{r}')$ is complex and provides a microscopic theory for the imaginary part of the optical potential (Vinh-Mau, 1977, Bonaccorso & Brink, 1982).

The resonating group method (RGM) is a microscopic theory for nucleus–nucleus scattering which takes into account the antisymmetry effects required by the Pauli principle. The RGM wave function for a single channel system of two nuclei with zero spin has the form

$$\mathscr{A}(\chi(\mathbf{r})\phi_1\phi_2) \tag{10.4}$$

Here $\phi_1$ and $\phi_2$ are internal wave functions of the two nuclei, $\chi(\mathbf{r})$ is the wave function of relative motion and $\mathscr{A}$ is the antisymmetrizer. The RGM equation for the wave function $\chi$ is

$$\int (h(\mathbf{r}, \mathbf{r}') - En(\mathbf{r}, \mathbf{r}'))\chi(\mathbf{r}')d\mathbf{r}' = 0 \tag{10.5}$$

where

$$\begin{Bmatrix} h(\boldsymbol{\rho}, \boldsymbol{\rho}') \\ n(\boldsymbol{\rho}, \boldsymbol{\rho}') \end{Bmatrix} = \left\langle \delta(\mathbf{r} - \boldsymbol{\rho})\phi_1\phi_2 \middle| \begin{Bmatrix} H - E_1 - E_2 \\ 1 \end{Bmatrix} \middle| \mathscr{A}(\delta(\mathbf{r} - \boldsymbol{\rho}')\phi_1\phi_2 \right\rangle \tag{10.6}$$

and $E_1, E_2$ are the energies of the two fragments calculated with $\phi_1, \phi_2$. The RGM equation (10.5) has a non-local hamiltonian and a non-local normalization operator $n(\mathbf{r}, \mathbf{r}')$.

The aim of the present chapter is to show how semi-classical methods can be applied to non-local potentials of the Feshbach type and to the RGM. The WKB method was first used by Austern (1965) to study non-local potentials in nuclear scattering problems. Recently the theory has been developed further by Horiuchi (1980*a*, *b*) and collaborators who have made many applications to RGM problems. Horiuchi's treatment of the subject is followed here. We show how non-local potentials with a short range $\beta$ of non-locality can be replaced first by momentum-dependent potentials and then by energy-dependent effective local potentials. Thus all the semi-classical methods developed in earlier chapters of this book can be applied to such non-local potentials.

The WKB method can only be used for certain classes of non-local potentials. The essential requirement is that $v(\mathbf{r}, \mathbf{r}')$ should be of short range in $\mathbf{r} - \mathbf{r}'$ and smooth and of longer range in $\mathbf{r} + \mathbf{r}'$. Horiuchi (1980*b*) shows that if $v(\mathbf{r}, \mathbf{r}')$ is of short range in $\mathbf{r} + \mathbf{r}'$ and longer range in $\mathbf{r} - \mathbf{r}'$ it can be replaced by an exchange potential and the semi-classical methods outlined in section 6.6 can be used. There are many examples of non-local potentials for which the WKB approximation is not valid. One example is a separable potential $v(\mathbf{r}, \mathbf{r}') = \lambda f(\mathbf{r}) f(\mathbf{r}')$. Such a potential has no classical analogue and semi-classical methods have no meaning for it.

## 10.2 The WKB approximation

We consider the Schrödinger equation

$$\int h(\mathbf{r}, \mathbf{r}')\psi(\mathbf{r}')d\mathbf{r}' = E\psi(\mathbf{r}) \tag{10.7}$$

where $h(\mathbf{r}, \mathbf{r}')$ is symmetric and rotationally invariant and consists of both local and non-local terms

$$h(\mathbf{r}, \mathbf{r}') = [-(\hbar^2/2\mu)\nabla_{\mathbf{r}}^2 + V_0(r)]\delta(\mathbf{r} - \mathbf{r}') + v(\mathbf{r}, \mathbf{r}') \tag{10.8}$$

Horiuchi (1980*a*, *b*) obtained the WKB approximation to (10.7) in terms of the Wigner transform $H(\mathbf{r}, \mathbf{p})$ of the non-local kernel

$$H(\mathbf{r}, \mathbf{p}) = \int h(\mathbf{r} - \tfrac{1}{2}\mathbf{s}, \mathbf{r} + \tfrac{1}{2}\mathbf{s}) \exp(i\mathbf{p}\cdot\mathbf{s}/\hbar)d\mathbf{s}$$

He shows that the results of the WKB approximation in section 8.3 generalize directly if

$$H(\mathbf{r}, \mathbf{p}) = \mathbf{p}^2/2\mu + V_0(r) + v_W(\mathbf{r}, \mathbf{p}) \tag{10.9}$$

is interpreted as a hamiltonian. In (10.9) $v_W(\mathbf{r}, \mathbf{p})$ is the Wigner transform of the non-local potential and it is a momentum-dependent potential. The approximations made in deriving the results in this section require that $v(\mathbf{r}, \mathbf{r}')$ has a short range of non-locality; that is, $v(\mathbf{r}, \mathbf{r} + \mathbf{s})$ has a short range $\beta$ in $\mathbf{s}$ and is a slowly varying function of $\mathbf{r}$. We obtain the WKB equations using a simplified version of Horiuchi's argument.

We seek a solution of equation (10.7) of the form used in section 8.3

$$\psi(\mathbf{r}) = A(\mathbf{r}) \exp(iW(\mathbf{r})/\hbar) \tag{10.10}$$

Because $h(\mathbf{r}, \mathbf{r} + \mathbf{s})$ has a short range in $\mathbf{s}$ and is a slowly varying function of $\mathbf{r}$ we make the following approximations

$$W(\mathbf{r} + \mathbf{s}) \simeq W(\mathbf{r}) + \mathbf{s}\cdot\nabla_{\mathbf{r}} W(\mathbf{r})$$
$$A(\mathbf{r} + \mathbf{s}) \simeq A(\mathbf{r}) + \mathbf{s}\cdot\nabla_{\mathbf{r}} A(\mathbf{r})$$
$$h(\mathbf{r}, \mathbf{r} + \mathbf{s}) \simeq (1 + \tfrac{1}{2}\mathbf{s}\cdot\nabla_{\mathbf{r}})h(\mathbf{r} - \tfrac{1}{2}\mathbf{s}, \mathbf{r} + \tfrac{1}{2}\mathbf{s})$$

when evaluating the integral in (10.7). Collecting the leading terms in powers of $\hbar$ we get

$$\int h(\mathbf{r}, \mathbf{r}+\mathbf{s})\psi(\mathbf{r}+\mathbf{s})d\mathbf{s}$$

$$\simeq \exp(iW(\mathbf{r})/\hbar)\left[A(\mathbf{r})H(\mathbf{r},\mathbf{p}) - i\hbar\left(\frac{\partial A}{\partial r_j} + \frac{1}{2}A\frac{\partial}{\partial r_j}\right)\frac{\partial H}{\partial p_j}\right]$$

$$= EA(\mathbf{r})\exp(iW(\mathbf{r})/\hbar) \tag{10.11}$$

where $\mathbf{p} = \nabla W$. The term independent of $\hbar$ in (10.11) gives

$$H(\mathbf{r}, \mathbf{p}(\mathbf{r})) = H(\mathbf{r}, \nabla W) = E \tag{10.12}$$

This is the Hamilton–Jacobi equation for the action function $W$ if $v_W(\mathbf{r},\mathbf{p})$ is interpreted as a momentum-dependent potential. It is a direct generalization of (8.28). The terms of order $\hbar$ give a continuity equation

$$\nabla\cdot(A^2\dot{\mathbf{r}}) = 0 \tag{10.13}$$

which is a generalization of (8.29). The velocity $\dot{\mathbf{r}}$ appears in (10.13) because $\dot{\mathbf{r}} = \partial H/\partial \mathbf{p}$ from Hamilton's equations. As in section 8.3 $A^2$ is a probability density and $\mathbf{j} = A^2\dot{\mathbf{r}}$ is a probability current.

The solution of (10.12) depends on the choice of boundary conditions. With boundary conditions ($\mathbf{p} \to \mathbf{p}_0$ as $t \to -\infty$) $W(\mathbf{r}, \mathbf{p})$ is given by the action integral (8.23) along a classical orbit with incident momentum $\mathbf{p}_0$ and final position $\mathbf{r}$. The normalization $A(\mathbf{r})$ is given by a van Vleck determinant (van Vleck, 1928)

$$A^2(\mathbf{r}) = \det\left(\frac{\partial^2 W}{\partial r_i \partial p_{0j}}\right) \tag{10.14}$$

If the non-local potential $v(\mathbf{r}, \mathbf{r}')$ is rotationally invariant its Wigner transform $v_W(r, p_r, \lambda)$ will be a function of the radial distance $r$ and momentum $p_r$, and the angular momentum $\lambda$. In a scattering problem the semi-classical phase $\delta(\lambda)$ is given by (8.31) where the radial momentum has to be found by solving the energy equation (10.12)

$$\frac{1}{2\mu}(p_r^2 + \hbar^2\lambda^2/r^2) + V_0(r) + v_W(r, p_r, \lambda) = E \tag{10.15}$$

If the contribution $\Delta$ of the non-local potential $v_W$ to the phase is calculated by the semi-classical perturbation formula then

$$\Delta\delta = -(1/2\hbar)\int_{-\infty}^{\infty} v_W(\mathbf{r}(t), \mathbf{p}(t))dt \tag{10.16}$$

where $\mathbf{r}(t), \mathbf{p}(t)$ is a classical orbit in the potential $V_0(r)$. This is the generalization of equation (3.61) to a non-local potential.

## 10.3 Equivalent local potentials

A non-local potential can always be replaced by a local potential which yields the same values for physical observables. Such a potential is called an equivalent local potential. A procedure for finding local potentials equivalent to certain non-local potentials was developed by Perey & Buck (1962). The method was generalized by Austern (1965) and Fiedeldey (1967). Austern's approach was based on a WKB approximation and in this section we discuss the method as described by Horiuchi (1980*a*, *b*). The equivalent local potential is defined by

$$\frac{1}{2\mu}(p_r^2 + \hbar^2\lambda^2/r^2) + V_{eq}(r, \lambda, E) = E \tag{10.17}$$

It is a function of position, angular momentum and energy and it is obtained by solving (10.15) for the radial momentum $p_r$ in terms of $r, \lambda, E$, and then substituting into (10.17). The WKB phase shifts calculated from (10.15) and from the equivalent local potential in (10.17) are identical. This is because the WKB phases depend on the radial momentum $p_r$ and by construction (10.15) and (10.17) give identical results. We denote the wave function for the equivalent local potential by $\psi^L(\mathbf{r})$. Austern (1965) and Fiedeldey (1967) show that $\psi$ and $\psi^L$ are proportional

$$\psi(\mathbf{r}) = F(\mathbf{r})\psi^L(\mathbf{r}) \tag{10.18}$$

The function $F(\mathbf{r})$ is usually called the Perey factor (Perey, 1963; Austern, 1965). Austern shows that WKB

$$F(r) = (1 - \partial V_{eq}/\partial E)^{\frac{1}{2}} \tag{10.19}$$

The origin of the factor is as follows. The WKB radial wave function (3.4) for a local potential contains a factor $(p_r)^{-\frac{1}{2}}$. In the case of a non-local or momentum-dependent potential this should be replaced by $(\mu\dot{r})^{-\frac{1}{2}}$ Hence

$$F(r) = (p_r/\mu\dot{r})^{\frac{1}{2}} = \left(\frac{\mu}{p_r}\frac{\partial H}{\partial p_r}\right)^{-\frac{1}{2}} \tag{10.20}$$

using Hamilton's equation for $r$. This can be reduced to (10.19) by using (10.15) and (10.17). The Perey factor goes to unity as $r \to \infty$ because $V_{eq}(r) \to 0$. Hence the wave functions $\psi$ and $\psi^L$ are identical for large $r$. In particular scattering phase shifts, which depend on the asymptotic form of the wave function, are not influenced by the Perey factor.

The equivalent local potential is a useful device because it is much easier to visualize than a non-local potential or even a momentum-dependent potential. Its definition in equation (10.17) is based on the WKB approximation and it may not be valid near classical turning points or in classically forbidden regions. In such cases $V_{eq}(r)$ can be obtained by Fiedeldey's (1967) theory.

## 10.4 The resonating group equation

The resonating group equation (10.5)

$$\int (h(\mathbf{r},\mathbf{r}') - En(\mathbf{r},\mathbf{r}'))\chi(\mathbf{r}')d\mathbf{r}' = 0 \qquad (10.21)$$

is even more complicated than the non-local Schrödinger equation (10.8) but the WKB method of section 10.2 can be applied to it (Horiuchi, 1983; Aoki & Horiuchi, 1982) provided that the kernels $h(\mathbf{r},\mathbf{r}')$ and $n(\mathbf{r},\mathbf{r}')$ have a short range of non-locality. If we put

$$\chi(\mathbf{r}) = A(\mathbf{r})\exp(iW(\mathbf{r})/\hbar)$$

and make the approximation (10.11) for $h$ and $n$ then we obtain the Hamilton–Jacobi equation

$$h_W(\mathbf{r},p) - En_W(\mathbf{r},\mathbf{p}) \qquad (10.22)$$

where $h_W$ and $n_W$ are the Wigner transforms of $h$ and $n$. Equation (10.22) can be written in the form (10.12) by defining $H(\mathbf{r},\mathbf{p}) = h_W(\mathbf{r},\mathbf{p})/n_W(\mathbf{r},\mathbf{p})$. The continuity equation equivalent to (10.13) is

$$\frac{\partial}{\partial r_j}\left[A^2(r)\frac{\partial}{\partial p_j}(h_W(\mathbf{r},\mathbf{p}) - En_W(\mathbf{r},\mathbf{p}))\right] = 0 \qquad (10.23)$$

This can also be written in the form

$$\nabla\cdot(A^2 n_W \dot{\mathbf{r}}) = 0$$

where $\dot{\mathbf{r}} = \partial H/\partial\mathbf{p}$. Comparing with (10.14) the normalization $A(\mathbf{r})$ is given by

$$A^2(\mathbf{r}) = \frac{1}{n_W}\det\left(\frac{\partial^2 W}{\partial r_i \partial p_{0j}}\right)$$

for a scattering problem. The action $W$ is again calculated from (8.23).

An effective local potential can be calculated from $H(\mathbf{r},\mathbf{p})$ in the same way as described in section 10.3. If $\chi^L(\mathbf{r})$ is the wave function calculated from the equivalent local potential then $\chi(\mathbf{r}) = F(\mathbf{r})\chi^L(\mathbf{r})$ where the Perey factor is given by

$$F(\mathbf{r}) = [(1 - \partial V_{eq}/\partial E)/n_W]^{\frac{1}{2}} \qquad (10.24)$$

## 10.5 Parity-dependent potentials

For a general non-local potential $v(\mathbf{r},\mathbf{r}')$ let $\beta_+$ denote the range in $\mathbf{r}+\mathbf{r}'$ and $\beta_-$ the range in $\mathbf{r}-\mathbf{r}'$. The methods of section 10.2 apply if $\beta_- \ll \beta_+$. We denote the local wave number by $k$. The condition for the WKB method of section 10.2 to be meaningful is $\beta_- \lesssim k^{-1} < \beta_+$. The condition $k^{-1} < \beta_+$ is the necessary short wave length condition for the validity of any WKB method. The other condition $\beta_- \lesssim k^{-1}$ must be satisfied for the non-local potential to contribute to the scattering. If $k^{-1} \ll \beta_-$ the WKB method is valid, but the wave function oscillates so rapidly that

$$\int v(\mathbf{r},\mathbf{r}')\psi(\mathbf{r}')d\mathbf{r}' \simeq 0$$

and the non-local potential is ineffective.

If the non-local potential has the property $\beta_+ \ll \beta_-$ and $\beta_+ \lesssim k^{-1} < \beta_-$ we must proceed in another way by writing

$$\int v(\mathbf{r},\mathbf{r}')\psi(\mathbf{r}')d\mathbf{r}' = \int v(\mathbf{r},-\mathbf{r}')P_{\mathrm{M}}\psi(\mathbf{r}')d\mathbf{r}'$$

where $P_{\mathrm{M}}$ is the Majorana exchange operator ($P_{\mathrm{M}}\psi(\mathbf{r}) = \psi(-\mathbf{r})$). Now the range of $v(\mathbf{r},-\mathbf{r}')$ in $\mathbf{r}+\mathbf{r}'$ is $\beta_-$ and in $\mathbf{r}-\mathbf{r}'$ it is $\beta_+$ and $v(\mathbf{r},-\mathbf{r}')$ can be treated by the WKB method. Hence $v(\mathbf{r},\mathbf{r}')$ can be replaced by a momentum-dependent exchange potential

$$v(\mathbf{r},\mathbf{r}') \rightarrow \bar{v}_{\mathrm{W}}(\mathbf{r},\mathbf{p})P_{\mathrm{M}} \tag{10.25}$$

where $\bar{v}_{\mathrm{W}}(\mathbf{r},\mathbf{p})$ is the Wigner transform of $v(\mathbf{r},-\mathbf{r}')$. The methods described in section 6.6 can be used to calculate the effects of this potential.

In a general case it may be possible to divide a non-local potential into a sum of two parts

$$v(\mathbf{r},\mathbf{r}') = v_{\mathrm{A}}(\mathbf{r},\mathbf{r}') + v_{\mathrm{B}}(\mathbf{r},\mathbf{r}') \tag{10.26}$$

where $v_{\mathrm{A}}$ has a short range in $\mathbf{r}-\mathbf{r}'$ and $v_{\mathrm{B}}$ has a short range in $\mathbf{r}+\mathbf{r}'$. Then $v(\mathbf{r},\mathbf{r}')$ is replaced by a direct part from $v_{\mathrm{A}}$ and an exchange part from $v_{\mathrm{B}}$. Aoki & Horiuchi (1982, 1983) have analysed the non-local kernels which occur in the RGM and find that a decomposition of the form (10.26) can indeed be made. If the mass numbers $A_1$ and $A_2$ of the two nuclei are very different they show that the terms of type B are negligible. If $A_1 \simeq A_2$ so that the two nuclei can exchange their identity by transfer of a few nucleons then the terms of type B are important. If $A_1 + A_2$ is large the kernels which give a significant contribution fall clearly either into class A or class B so that a description of the scattering in terms of direct and exchange potentials is valid. Aoki and Horiuchi also find that the equivalent local potentials are deep and resemble the folding potentials (2.8) discussed in section 2.2.

# 11
# Fusion reactions

## 11.1 Classical theory

In a fusion reaction two nuclei combine together to form a composite system which is characterized by its total mass, charge, energy and angular momentum. It corresponds to Bohr's classical compound nucleus if it has reached equilibrium with respect to all other internal degrees of freedom. The most convincing experimental evidence for fusion is provided by the observation of evaporation residues. These are reaction products with masses near those of the compound system. They are formed when the compound nucleus evaporates a few neutrons or protons or when it emits $\gamma$-rays on its way to the detector. Fusion may also be followed by fission. The contribution of this process will be missed out if fusion cross-sections are measured by observing evaporation residues.

Work on fusion reactions with light targets and projectiles has been reviewed by Bass (1980). There is also an extensive review by Vaz, Alexander & Satchler (1981). Recent advances in the field have been described by Broglia (1983).

Fusion is expected to be associated with the lowest partial waves. In the simplest classical theory there is a critical angular momentum $\lambda_f$ at each energy. The fusion probability is unity for angular momenta $\lambda < \lambda_f$ and zero for $\lambda > \lambda_f$. This sharp cut-off model gives a fusion cross-section

$$\sigma_f = \pi\lambda_f^2/k^2 = \pi b_f^2 \tag{11.1}$$

where $b_f = \lambda_f/k$ is a critical impact parameter. Fusion cross-sections can be estimated by assuming that the distance of closest approach in a Rutherford orbit with impact parameter $b_f$ is equal to the strong absorption radius $R_s$. Using equation (2.1)

$$b_f^2 = R_s^2 - 2a_C R_s$$

where $a_C$ is the Coulomb length parameter (1.2). Substituting in (11.1) and using (1.2) gives

$$\begin{aligned} \sigma_f &\simeq \pi R_s^2(1 - V_{Cs}/E); \quad E > V_{Cs} \\ &= 0; \qquad\qquad\qquad\quad E < V_{Cs} \end{aligned} \tag{11.2}$$

where $V_{Cs} = Z_1 Z_2 e^2/R_s$ is the Coulomb interaction potential at the strong absorption radius, and E is the centre-of-mass energy. At high energies (11.2) overestimates the fusion cross-section because not all reactions lead to fusion. Below the Coulomb barrier fusion is classically forbidden and equation (11.2) gives a zero cross-section. Nevertheless fusion can occur at subbarrier energies because of barrier penetration.

The quantal formula for a fusion cross-section is

$$\sigma_f = (\pi/k^2)\sum_l (2l+1)P_l \tag{11.3}$$

where the transmission coefficient $P_l$ is the probability of fusion for an incident partial wave $l$. Equation (11.3) is exact if the initial nuclei are different and have zero spin. It is also valid for non-zero spins if the target-projectile interaction is spin-independent. A sharp cut-off model with

$$P_l = 1; \quad l \leqslant l_f$$
$$P_l = 0; \quad l > l_f$$

gives

$$\sigma_f = (\pi/k^2)(l_f + 1)^2 \tag{11.4}$$

Equation (11.4) is equivalent to the classical formula (11.1) when $l_f$ is large.

Equation (11.3) must be modified if the target and projectile are identical since wave functions with the proper exchange symmetry must be used. If the identical nuclei have ground state spins $I$ then (11.3) is replaced by

$$\sigma_f = (\pi/k^2)\sum_l (2l+1)P_l(1 + (-1)^{2I+l}/(2I+1)) \tag{11.5}$$

where $P_l$ is assumed to be spin-independent.

## 11.2 Sub-barrier fusion

Until recently most experimental studies of sub-barrier fusion have been made in light systems where the barriers are within reach of smaller accelerators. Now higher energy tandem accelerators are available, new techniques are being developed for detecting evaporation residues and the range of pairs of nuclei investigated is being extended. The optical model has been successful in describing fusion cross-sections for light systems. In these applications the $P_l$ in equation (11.3) are replaced by optical model transmission coefficients and the fusion cross-section is identified with the total reaction cross-section. This assumption is justified for most light systems except for special cases like $^{14}N + ^{14}N$ at low energies where the neutron transfer is large. Many optical potentials have been used and one particular Woods–Saxon potential deduced by Reeves (1966) reproduces fusion cross-sections for many light systems quite well. The parameters of this potential are $V = 50\,\text{MeV}$, $W = 10\,\text{MeV}$, $r_v = 1.26\,\text{fm}$, $r_w = 1.215\,\text{fm}$, $a_v = 0.44\,\text{fm}$, $a_w = 0.45\,\text{fm}$. The radius parameter of the real potential (2.10) is given by $R_v = r_v(A_1^{\frac{1}{3}} + A_2^{\frac{1}{3}})$ and similarly for $R_w$.

The physical significance of the imaginary part of the optical potential for large separations is not clear for fusion calculations. Alternative models have been proposed which use a real potential and assume that fusion occurs when the nuclear density distributions begin to overlap. The ingoing wave boundary condition model of Christensen & Switkowski (1977) uses the requirement that the wave function has an ingoing character at a suitably chosen contact radius. The average non-resonant behaviour of many fusion cross-sections can be reproduced well with real potentials which fit elastic scattering data.

Cujec & Barnes (1976) have used an even simpler approach. They approximate the Coulomb barrier by an inverted parabola and use the transmission coefficient formula of Hill & Wheeler (1953),

$$P_l = \frac{1}{1 + \exp((V_{Bl} - E)/\Delta E)} \tag{11.6}$$

In equation (11.6)

$$\Delta E = \hbar\omega_B/2\pi \tag{11.7}$$

$$\omega_B^2 = \frac{1}{\mu}\left[\frac{\partial^2 V}{\partial r^2}\right]_B \tag{11.8}$$

$$V_{Bl} = V(r_B) + \frac{\hbar^2 l(l+1)}{2\mu r_B^2} \tag{11.9}$$

and $V(r)$ is the Coulomb plus nuclear potential. The quantity $V_{Bl}$ is the barrier height in the partial wave $l$ and $\Delta E$ is a measure of the barrier thickness. Equation (11.6) for $P_l$ contains some simplifications because the barrier radius $r_B$ and thickness $\Delta E$ are assumed to be independent of $l$. With these assumptions the sum (11.3) can be evaluated to a very good approximation by using the leading term of the Poisson sum formula to give a closed formula for the fusion cross-section (Wong, 1973)

$$\sigma_f = \pi r_B^2(\Delta E/E)\ln[1 + \exp((E - V_B)/\Delta E)] \tag{11.10}$$

where $V_B = V(r)$. Table (11.1) shows some typical values of the parameters $r_B$, $V_B$ and $\Delta E$ for several systems which have been studied experimentally. They were calculated using the potential (2.16) of Christensen & Winther (1976). It is noteworthy that the parameter $\Delta E$ which measures the barrier

Table 11.1. *Typical fusion barrier parameters*

| Nuclei | $r_B$(fm) | $V_B$(MeV) | $\Delta E$ | $l_B$ |
|---|---|---|---|---|
| $^{16}O + ^{16}O$ | 8.1 | 10.4 | 0.51 | 1.0 |
| $^{58}Ni + ^{58}Ni$ | 10.8 | 98.0 | 0.72 | 5.0 |
| $^{16}O + ^{148}Sm$ | 11.0 | 61.0 | 0.80 | 3.6 |
| $^{40}Ar + ^{122}Sn$ | 11.6 | 105.0 | 0.72 | 5.4 |

Barrier radius ($r_B$), barrier height ($V_B$), $\Delta E = \hbar\omega_B/2\pi$, number of contributing partial waves ($l_B$).

thickness does not change much from system to system. The parabolic approximation and equation (11.10) derived from it become inaccurate when $E \ll V_B$.

The number $l_B$ of partial waves which contribute significantly to the sub-barrier fusion cross-section can be estimated from

$$\hbar^2 l_B(l_B + 1)/(2\mu r_B^2) = \Delta E$$

This number is shown in the last column of table 11.1. and can be quite large for heavier systems. If $n_B$ is the Sommerfeld parameter for a centre-of-mass energy $E = V_B$ then

$$l_B/n_B \simeq 2(\Delta E/V_B)^{\frac{1}{2}}$$

This ratio is always small.

When the incident energy is well above the top of the Coulomb barrier ($E - V_B \gtrsim 2\Delta E$) equation (11.10) simplifies to give

$$\sigma_f \simeq \pi r_B^2(1 - V_B/E) \tag{11.11}$$

Equation (11.11) has the same form as the classical fusion cross-section (11.2). For incident energies well below the barrier ($V_B - E \gtrsim 2\Delta E$)

$$\sigma_f \simeq \pi r_B^2(\Delta E/E)\exp((E - V_B)/\Delta E) \tag{11.12}$$

and the fusion cross-section decreases exponentially. In fact, the parabolic approximation is inaccurate when $E \ll V_B$ and (11.12) overestimates the fusion cross-section.

## 11.3 Influence of intrinsic degrees of freedom

Recent experiments on fusion with heavy targets and projectiles show that the simple barrier penetration model discussed in section 11.2 cannot account for the energy variation of the fusion cross-section for energies near and below the Coulomb barrier in such systems. Stokstad *et al.* (1978) studied the excitation function for fusion of $^{16}O$ with different samarium isotopes. The results show a large variation of the cross-section with mass number at the lowest energy investigated. Above the barrier, on the other hand, the cross-sections become almost identical. This behaviour gives clear evidence of a direct influence of nuclear deformation. The effect can be understood qualitatively by noting that the nuclear interaction is stronger and the Coulomb barrier lower when a prolate deformed nucleus is hit by a spherical projectile near its poles than when it is hit near its equator. The total fusion cross-section corresponds to an average over all orientations of the nuclear symmetry axis. At sub-barrier energies the relation between the effective barrier height and the cross-section is very non-linear and the cross-section will be dominated by contributions from those orientations which give the largest transmission coefficient. The net effect is a strong increase in fusion cross-section with increasing deformation.

The same kind of effect is important for fusion cross-sections of $^{40}Ar$ on tin isotopes (Reissdorf *et al.*, 1982). Recently another kind of effect has been found by Beckerman *et al.* (1980, 1983). Some of the results are shown in fig. 11.1. The diagram shows fusion cross-sections for (*a*) $^{58}Ni + {}^{58}Ni$, (*b*) $^{64}Ni + {}^{64}Ni$ and (*c*) $^{58}Ni + {}^{64}Ni$. The energy variations in cases (*a*) and (*b*) are very similar: The difference can be accounted for by a shift of about 4.4 MeV in the barrier height. In case (*c*) the cross-section has a qualitatively different variation with energy and is relatively large at energies of several MeV below the barrier. This difference cannot be explained by deformation effects. Broglia *et al.* (1983) have shown that two-neutron transfer effects can explain the difference.

The common feature of all these examples is that they show the effects of intrinsic degrees of freedom on the quantum tunnelling of a collective variable. Several authors have discussed this problem (Carlson & Hussein (1982), Broglia *et al.* (1983), Dasso *et al.* (1983*a*, 1983*b*), Jacob & Smilansky (1983), Lindsay & Rowley (1984)). The general conclusion reached by these authors is that the heavy-ion fusion cross-section is given by

$$\sigma_f(E) = \sum_i q_i \sigma_W(E - \varepsilon_i) \tag{11.13}$$

where $\sum_i q_i = 1$ and $\sigma_W(E - \varepsilon_i)$ are fusion cross-sections calculated from a simple barrier penetration formula such as the Wong formula (11.10). The interpretation of (11.13) is that fusion occurs through certain eigenchannels or transition states labelled by *i*. The transition state *i* has an energy $\varepsilon_i$ and the effective barrier height in that channel is $V_B + \varepsilon_i$. The quantity $q_i$ is the

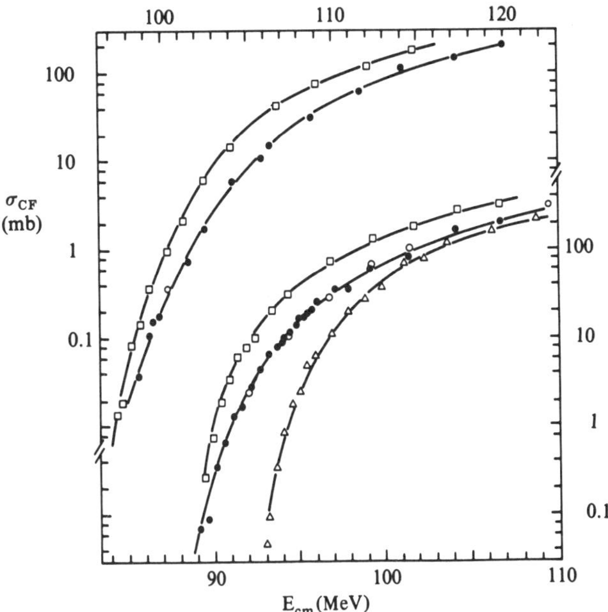


Fig. 11.1. Excitation functions for complete fusion. Constructed using weighted, average centre-of-mass energies. Lower portion: $^{58}Ni + ^{58}Ni$ (open triangles), $^{58}Ni + ^{64}Ni$ (filled circles), $^{64}Ni + ^{58}Ni$ (open circles), and $^{64}Ni + ^{64}Ni$ (open squares). Upper portion: $^{58}Ni + ^{74}Ge$ (filled circles) and $^{64}Ni + ^{74}Ge$ (open squares). Smooth curves drawn through the data points are visual guides (from Beckerman *et al.*, 1983).

probability of arriving at the barrier in the transition state $i$ and $\sigma_W(E - \varepsilon_i)$ is the probability of fusion in that state.

When the incident energy is well above the top of the barrier then $\sigma_W(E)$ is a slowly varying function of $E$ and (11.13) gives

$$\sigma_f(E) \approx \sigma_W(E - \bar{\varepsilon})$$

where $\bar{\varepsilon} = \sum_i q_i \varepsilon_i$ is an average of the transition state energies. For sub-barrier fusion the term in equation (11.13) with the smallest energy $\varepsilon_0$ gives the dominant contribution and

$$\sigma_f(E) \approx q_0 \sigma_W(E - \varepsilon_0)$$

Thus the energy width of the transition region is approximately $\bar{\varepsilon} - \varepsilon_0$. Deformations enhance the sub-barrier fusion cross-section. Then the barrier height depends on deformation and

$$\bar{\varepsilon} - \varepsilon_0 \approx \tfrac{1}{2}(V_B(\max) - V_B(\min))$$

The existence of a neutron transfer channel with a negative $Q$ would give $\bar{\varepsilon}-\varepsilon_0 \approx |Q|$ and would produce an enhancement of the sub-barrier cross-section.

Equation (11.13) can be derived from a variety of starting points. The most direct approach to the fusion problem is to make a full quantal coupled-channels calculation. Broglia *et al.* (1983) and Dasso *et al.* (1983*a, b*) use a very simplified coupled channels model in which the intrinsic states are degenerate and the coupling can be removed by a unitary transformation. Lindsay & Rowley (1984) also use this kind of model but discuss angular momentum coupling effects explicitly. Jacob & Smilansky (1983) and Takigawa & Bertsch (1984) use a path integral method based on influence functionals. In the following section we derive equation (11.13) from a path integral approach. The essential approximation which we use is that the tunnelling is not treated in a self-consistent way. The influence functional method gives a better approximation but is also more complicated. The qualitative features of the results are very similar.

The path integral method described in section 11.4 is equivalent to a version of the WKB method given by Brink, Nemes & Vautherin (1983) for the coupling of a collective variable $r$ to intrinsic degrees of freedom $\xi$. This method assumes a trial wave function of the form

$$\psi(r,\xi)=\phi(r,\xi)\exp(iW(r)/\hbar)$$

## 11.4 A path integral for fusion

We consider a system with hamiltonian

$$H=-\frac{\hbar^2}{2\mu}\frac{\partial^2}{\partial r^2}+V_0(r)+h(r,\xi) \tag{11.14}$$

$$h(r,\xi)=H_0(\xi)+v(r,\xi) \tag{11.15}$$

Here the radial coordinate $r$ is the separation of the centres of the interacting nuclei and $V_0(r)$ is the Coulomb plus a real nuclear potential. The tunnelling coordinate $r$ is coupled to the intrinsic coordinates $\xi$ by the interaction $v(r,\xi)$ and $H_0(\xi)$ is the hamiltonian of the intrinsic variables. By using the hamiltonian (11.14) to calculate fusion we assume that the relative motion is along the radial direction. This is approximately correct for fusion with orbital angular momentum $l=0$. An extension to the case of non-zero $l$ will be discussed later in the section.

In order to study tunnelling we write a Pechukas path integral (9.12) for the amplitude $K_{\beta\alpha}(r_1,r_0,T)$ so that the system moves from an initial position $r_0$ at $t = 0$ to a final position $r_1$ at $t = T$, while the internal state changes from $\alpha$ to $\beta$,

$$K_{\beta\alpha}(r_1, r_0, T) = \int D[r(t)] T_{\beta\alpha}[r(t)] \exp((i/\hbar)S_0)[r(t)] \tag{11.16}$$

where $S_0[r(t)]$ is the action

$$S_0[r(t)] = \int_0^T (\tfrac{1}{2}\mu\dot{r}^2 - V_0(r))dt \tag{11.17}$$

The factor $T_{\beta\alpha}[r(t)]$ is the amplitude for a transition between the intrinsic eigenstates $\alpha$ and $\beta$ and the relative coordinate passes along $r(t)$. This amplitude is calculated from the time-dependent coupled equations (9.6) and (9.7). The integral (11.16) is over all paths satisfying the boundary conditions

$$r(0) = r_0, \quad r(T) = r_1 \tag{11.18}$$

We evaluate the path integral (11.16) using the approximation developed in section 9.3. The stationary path is determined by the condition $\delta S_0[r(t)] = 0$ and is a classical path $r_0(t)$ in the potential $V_0(r)$ which satisfies the boundary conditions (11.18). The factor $T_{\beta\alpha}[r(t)]$ is assumed to be a slowly varying functional of the path. It is evaluated for the stationary path $r_0(t)$ and taken outside the integral. The resulting formula for the propagator is

$$K_{\beta\alpha}(\mathbf{r}_1, \mathbf{r}_0, T) \approx T_{\beta\alpha}[\mathbf{r}_0(t)] K_0(\mathbf{r}_1, \mathbf{r}_0, T) \tag{11.19}$$

where $K_0$ is the semi-classical propagator (7.19) for the potential $V_0(r)$.

The next step is to evaluate the energy propagator (section 7.4)

$$G_{\beta\alpha}(r_1, r_0, E) = (1/i\hbar) \int_0^\infty dT K_{\beta\alpha}(r_1, r_0, T) \exp(iET/\hbar) \tag{11.20}$$

$$= (1/i\hbar) \int_0^\infty dT K_0(r_1, r_0, T) T_{\beta\alpha}[r_0(t)] \exp(iET/\hbar) \tag{11.21}$$

To obtain the fusion probability we must evaluate the energy propagator for an initial separation $r_0$, which is large, to a final separation $r_1$, which is inside the Coulomb barrier formed by $V_0(r)$. If the barrier height is $V_\mathrm{B}$ we can distinguish three cases:

(i) $E \gg V_\mathrm{B}$: when the incident energy is well above the barrier we evaluate the time integral in equation (11.21) by using the stationary phase approximation as in section 7.4 and by assuming that $T_{\beta\alpha}[r_0(t)]$ is a slowly varying function of $T$. The stationary path $r_\mathrm{E}(t)$ is the classical path with energy $E$ which starts at $r_0$ and ends at $r_1$. The stationary time $T_\mathrm{s}(E)$ is the time taken to move from $r_0$ to $r_1$ in the potential $V_0(r)$. The stationary phase formula is

$$G_{\beta\alpha}(r_1, r_0, E) \approx T_{\beta\alpha}[r_E(t)]G_0(r_1, r_0, E) \tag{11.22}$$

In this equation (11.22) $G_0$ is the semi-classical propagator for the relative motion in the potential $V_0(r)$ and is given by equation (7.37). The final intrinsic state is not observed in a fusion reaction and the fusion probability is proportional to

$$\sum_\beta |G_{\beta\alpha}(r_1, r_0, E)|^2 = \sum_\beta |T_{\beta\alpha}[r_E(t)]|^2 |G_0(r_1, r_0, E)|^2 \tag{11.23}$$

In our approximation all the effects of the coupling are contained in the transition amplitudes but they do not influence the final result because $T_{\beta\alpha}$ is a matrix element (9.6) of a unitary operator and $\sum_\beta |T_{\beta\alpha}|^2 = 1$. The fusion probability is $P_0 = 1$.

(ii) $E \ll V_B$: when the incident energy is well below the top of the barrier the time integral in (11.21) can be evaluated by the saddle point method as was done in section 7.8. The result is still given by equation (11.22) but there are some important differences. The first is that $G_0(r_1, r_0, E)$, which is given by equation (7.58), contains a barrier penetration factor $\exp(-Q(E))$. The second difference is that the transition amplitude $T_{\beta\alpha}$ is not unitary. The reason is that the stationary time is complex

$$T_S = T_0 - i\tau. \tag{11.24}$$

Here $T_0$ is the real time for propagation from $r_0$ up to the nearside of the barrier and from the farside of the barrier to $r_1$, while $-i\tau$ is the imaginary time for propagation through the barrier. It is given by the integral (7.57). The non-unitary character of $T_{\beta\alpha}$ can be seen most clearly if the intrinsic hamiltonian (11.15) including the effects of coupling is independent of $r$ in the barrier region

$$h(r, \xi) = h(r_B, \xi) = h_B(\xi)$$

Let $\phi_i$ be the eigenstates of $h_B$

$$h_B\phi_i(\xi) = \varepsilon_i\phi_i(\xi) \tag{11.25}$$

and $\varepsilon_i$ be the corresponding eigenvalues. We choose the ground state energy of the internal hamiltonian $H_0(\xi)$ to be zero. Hence the lowest eigenvalue $\varepsilon_0$ of $h_B$ is a measure of the energy shift due to the interaction. The sign and magnitude of the shift depend on the nature of the intrinsic variables and of the interaction. The $\phi_i$ are the transition states referred to in section 11.3. The coupled equations (9.6), (9.7) for $T_{\beta\alpha}$ can be solved in this case and

$$T_{\beta\alpha}[r_E(t)] = \sum_i B_{\beta i} \exp(-\varepsilon_i\tau/\hbar)A_{i\alpha} \tag{11.26}$$

Here $A_{i\alpha}$ is a transition amplitude from the initial intrinsic state $\alpha$ to the transition state $i$, as the projectile comes up to the barrier. The exponential in (11.26) describes propagation through the barrier and $B_{\beta i}$ is a transition amplitude from the state $i$ to the final state $\beta$ after passage through the barrier. Both the matrices $A_{i\alpha}$ and $B_{\beta i}$ are unitary. Non-unitary effects due to barrier penetration are contained in the exponential factor in (11.26). Because $B_{\beta i}$ is unitary

$$\sum_{\beta}|T_{\beta\alpha}|^2 = \sum_{i}|A_{i\alpha}|^2 \exp(-2\varepsilon_i\tau/\hbar) \tag{11.27}$$

The total fusion probability calculated from (11.23) is

$$P_0 = \exp(-2Q(E))\sum_{i}|A_{i\alpha}|^2 \exp(-2\varepsilon_i\tau/\hbar) \tag{11.28}$$

The barrier penetration integral

$$Q(E) = \int_a^b [(2\mu/\hbar^2)(V_0(r) - E)]^{\frac{1}{2}} dr \tag{11.29}$$

where $a$ and $b$ are the turning points on the opposite sides of the barrier. For a parabolic barrier

$$Q(E) = \pi(V_{\mathrm{B}} - E)/\hbar\omega_{\mathrm{B}} \tag{11.30}$$

The first factor in (11.28) comes from $|G_0|^2$ in equation (11.23) and is the fusion probability in the absence of coupling. The second factor contains the effects of the coupling. The contribution of a transition state $\phi_i$ with a high energy $\varepsilon_i$ is reduced because of the exponential factor in (11.28). If the energies $\varepsilon_i$ are not too large

$$Q(E) + \varepsilon_i\tau/\hbar \simeq Q(E - \varepsilon_i)$$

because $\tau/\hbar = -dQ/dE$, and equation (11.28) can be written as

$$P = \sum_{i}|A_{i\alpha}|^2 \exp[-2Q(E - \varepsilon_i)] \tag{11.31}$$

Equation (11.31) has the same structure as (11.13). The factor $|A_{i\alpha}|^2$ is the probability of reaching the transition state $\phi_i$ from the initial state $\alpha$ as the projectile comes up to the barrier and can be identified with $q_i$ in (11.13). The second factor $\exp(-2Q(E - \varepsilon_i)$ is the probability of barrier penetration in the transition state.

(iii) $E \simeq V_{\mathrm{B}}$: when the incident energy is near the top of the barrier the above argument cannot be used. This is because the stationary time $T_{\mathrm{s}}$ for the passage from $r_0$ to $r_1$ (7.59) has a logarithmic divergence when $E \to V_{\mathrm{B}}$. When $E$ is near $V_{\mathrm{B}}$ the velocity is very slow near the barrier and the projectile takes a long time to cross it. In section 7.6 the time integral in the formula for $G$ when $E \simeq V_B$ was evaluated by a uniform approximation. A similar method can be used here.

The transition amplitude $T_{\beta\alpha}[r_0(t)]$ in the integral (11.21) for $G_{\beta\alpha}$ is calculated for a classical path $r_0(t)$ which passes from $r_0$ to $r_1$ in a time $T$. This path also has a definite energy $E(T)$ which is a function of $T$. First we study the dependence of $T_{\beta\alpha}$ on $E$ or equivalently on $T$. Let $h_B(\xi) = h_B(r, \xi)$ be the intrinsic hamiltonian (11.15) including the coupling when the relative motion is at the barrier position $r = r_B$ and define transition states and energies as in equation (11.25). Next we make a decomposition similar to (11.26)

$$T_{\beta\alpha}[r_0(t)] = \sum_i b_{\beta i} \exp(-i\varepsilon_i T/\hbar) a_{i\alpha} \tag{11.32}$$

Here $a_{i\alpha}$ is an amplitude for a transition from the initial state $\alpha$ at $r_0$ to the transition state $\phi_i$ at $r_B$. It is calculated in the interaction representation with respect to $h_B$. Similarly $b_{\beta i}$ is an amplitude for a transition from $\phi_i$ at $r_B$ to $\beta$ at $r_1$. Now all the singular behaviour of $T_{\beta\alpha}$ as $E \to V_B$ is contained in the exponential factor in (11.32). The amplitudes $b_{\beta i}$ and $a_{i\alpha}$ are slowly varying functions of $E$ and to a good approximation can be replaced by their values at $E = V_B$. Then (11.21) becomes

$$G_{\beta\alpha}(r_1, r_0, E) = \sum_i b_{\beta i} a_{i\alpha}/(i\hbar) \int_0^\infty dT \exp((E - \varepsilon_i)T/\hbar) K_0(r_1, r_0, T)$$

$$= \sum_i b_{\beta i} a_{i\alpha} G_0(r_1, r_0, E - \varepsilon_i) \tag{11.33}$$

In equation (11.33) $G_0(r_1, r_0, E - \varepsilon_i)$ is the propagator in the energy representation (7.66) calculated by the uniform approximation. At the top of the barrier $b_{\beta i}$ and $a_{i\alpha}$ are unitary matrices and (11.33) yields

$$\sum_\beta |G_{\beta\alpha}(r_1, r_0, E)|^2 = \sum_i |a_{i\alpha}|^2 |G_0(r_1, r_0, E - \varepsilon_i)|^2 \tag{11.34}$$

This equation leads to a fusion probability

$$P_0 = \sum_i |a_{i\alpha}|^2 / [1 + \exp(2Q(E - \varepsilon_i))] \tag{11.35}$$

This expression is a direct generalization of (11.31). The improved form for the penetration coefficient comes from the uniform approximation (7.66) to $G_0(r_1, r_0, E)$.

Equations (11.31) and (11.35) give approximate expressions for the fusion probability in a partial wave $l = 0$. When the $z$-axis is parallel to the incident direction then the $z$-component of the intrinsic angular momentum is conserved during the fusion process. If the initial state $\alpha$ has zero angular momentum the amplitude $a_{i\alpha}$ is non-zero only for transition states $\phi_i$ with angular momentum component $M_i = 0$.

In order to evaluate the fusion cross-section we need the fusion probability for $l \neq 0$. The simplest assumption is that the main effect comes from the centrifugal potential. According to equations (11.31) and (11.35) the fusion probability $P_0(E - V_B)$ is a function of the energy difference $(E - V_B)$ between the incident energy and the unperturbed barrier energy $V_B$. If $l \neq 0$ the effect of the centrifugal barrier would give

$$P_l = P_0(E - V_B - \hbar^2 l(l+1)/(2\mu r_B^2)) \tag{11.36}$$

When the transition probability $|a_{i\alpha}|^2$ is energy-independent then (11.36) leads directly to (11.13) with $q_i = |a_{i\alpha}|^2$. Lindsay & Rowley (1984) are able to discuss angular momentum effects explicitly in their model of fusion. They arrive at a result which is essentially equivalent to (11.36).

We conclude this section with some remarks on the transition probabilities $|A_{i\alpha}|^2$ or $|a_{i\alpha}|^2$ from the initial intrinsic state $\alpha$ to the transition state $\phi_i$ at the barrier. The calculation of these quantities depends on a detailed model of the intrinsic hamiltonian $H_0(\xi)$ and the coupling $v(r, \xi)$ but we can make some general comments in two limiting cases. The first is an adiabatic approximation. If the approach to the barrier is slow the intrinsic state has time to adjust during the approach phase of the reaction. The nuclei start off in their ground states and will tend to be in the transition state $\phi_0$ with lowest energy at the barrier. Hence $|a_{0\alpha}|^2 \simeq 1$ and

$$P_0 \approx [1 + \exp(2Q(E - \varepsilon_0))]^{-1}$$

The system fuses in the transition state of lowest energy. In this limit the effect of the coupling is just to shift the barrier. The energy dependence of the fusion cross-section near the barrier will be the same as predicted by the simple theory of section (11.2). This does not account for the experimental data in heavier systems.

The second limiting case is a sudden approximation. If the approach to the barrier is fast the intrinsic state does not have time to change during the approach phase. Then, except for a phase, $A_{i\alpha}$ is equal to the overlap $\langle \phi_i | \alpha \rangle$ between the transition state $\phi_i$ and the initial internal state $|\alpha\rangle$. This is the limiting case which has been considered by most authors. The true situation is somewhat in between and a few low energy intrinsic states will be excited in the approach phase.

APPENDIX A

# Properties of special functions

Some properties of special functions which have been used in earlier chapters are collected in this appendix. Many references are made to Abramowitz & Stegun (1965). These references will be abbreviated by 'AS' followed by a section number.

## A1 The gamma-function

An asymptotic formula for the gamma-function is given by (AS, 6.1.41)

$$\ln\Gamma(z) \sim (z-\tfrac{1}{2})\ln z - z + \tfrac{1}{2}\ln(2\pi) + \cdots z\to\infty, |\arg z| < \pi \tag{A.1}$$

A special case which appears in barrier penetration amplitudes is

$$\Gamma(\tfrac{1}{2}+i\beta) \sim (2\pi)^{\frac{1}{2}} \exp\{i(\beta\ln\beta - \beta) - \tfrac{1}{2}\pi\beta\} \tag{A.2}$$

The Coulomb phase is

$$\begin{aligned}\sigma_l &= \arg\Gamma(l+1+in) = \operatorname{Im}\ln\Gamma(l+1+in)\\ &\sim \tfrac{1}{2}n\ln(n^2+\lambda^2) - n + \lambda\tan^{-1}(n/\lambda)\end{aligned} \tag{A.3}$$

$(\lambda = l+\frac{1}{2})$ which holds if $n$ is large or if $l$ is large. If $\beta$ is real (AS, 6.1.30) gives

$$|\Gamma(\tfrac{1}{2}+i\beta)|^2 = \pi/\cosh(\pi\beta) \tag{A.4}$$

## A2 Airy functions

The Airy function appears in derivations of connection formulae and in uniform approximations. The functions Ai($z$) and Bi($z$) satisfy the differential equation (AS, 10, 4.1)

$$\omega'' - z\omega = 0$$

They can be calculated from a power series (AS, 10.4.2). Asymptotic expansions for large $|z|$ are given by (AS, 10.4.59–68). In the following $\zeta = \frac{2}{3}z^{\frac{3}{2}}$.

When

$$|\arg z| < \pi$$

$$\mathrm{Ai}(z) \sim \tfrac{1}{2}\pi^{-\frac{1}{2}}z^{-\frac{1}{4}}\exp(-\zeta) \tag{A.5}$$

$$\mathrm{A'i}(z) \sim -\tfrac{1}{2}\pi^{-\frac{1}{2}}z^{\frac{1}{4}}\exp(-\zeta) \tag{A.6}$$

When

$$|\arg z| < \tfrac{2}{3}\pi$$

$$\mathrm{Ai}(-z) \sim \pi^{-\frac{1}{2}} z^{-\frac{1}{4}} \sin(\zeta + \tfrac{1}{4}\pi) \tag{A.7}$$

$$\mathrm{A'i}(-z) \sim -\pi^{-\frac{1}{2}} z^{\frac{1}{4}} \cos(\zeta + \tfrac{1}{4}\pi) \tag{A.8}$$

$$z^{\frac{1}{4}}\mathrm{Ai}(-z) \pm iz^{-\frac{1}{4}}\mathrm{A'i}(-z) \sim \pi^{-\frac{1}{2}} \exp[\pm i(\zeta - \tfrac{1}{4}\pi)] \tag{A.9}$$

## A3 Bessel functions

Bessel functions occur in asymptotic approximations for Legendre functions and spherical harmonics. We have used the following integral representations (AS, 9.1.21)

$$J_n(z) = (i^{-n}/\pi) \int_0^\pi \exp(iz\cos\phi)\cos(n\phi)d\phi \tag{A.10}$$

$$= (1/\pi) \int_0^\pi \cos(z\sin\phi - n\phi)d\phi \tag{A.11}$$

where $n$ is an integer. The following representation for $K_\nu(z)$ is also useful,

$$K_\nu(z) = \int_0^\infty \exp(-z\cosh t)\cosh(\nu t)dt; \ |\arg z| < \tfrac{1}{2}\pi \tag{A.12}$$

The following symmetry properties follow from (A.11),

$$J_{-n}(z) = J_n(-z) = (-1)^n J_n(z) \tag{A.13}$$

There are many recurrence relations (AS, 9.1.27). We have used one of them

$$\left.\begin{aligned} J_n'(z) &= \tfrac{1}{2}(J_{n-1}(z) - J_{n+1}(z)) \\ J_0'(z) &= -J_1(z) \end{aligned}\right\} \tag{A.14}$$

We have used the asymptotic formulae which are valid for large $|z|$

$$J_n(z) \sim \left(\frac{2}{\pi z}\right)^{\frac{1}{2}} \cos(z - \tfrac{1}{2}m\pi - \tfrac{1}{4}\pi) \tag{A.15}$$

$$J_n'(z) \sim -\left(\frac{2}{\pi z}\right)^{\frac{1}{2}} \sin(z - \tfrac{1}{2}m\pi - \tfrac{1}{4}\pi) \tag{A.16}$$

$$(J_n(z) \pm iJ_n'(z)) \sim \left(\frac{2}{\pi z}\right)^{\frac{1}{2}} i^{\pm n} \exp\{\mp i(z - \tfrac{1}{4}\pi)\} \tag{A.17}$$

## A4 Legendre polynomials and spherical harmonics

Some properties of Legendre functions

$$P_l(-x) = (-1)^l P_l(x) \tag{A.18}$$

Recurrence relation (AS, 8.5.3)

$$(l+1)P_{l+1}(x) = (2l+1)xP_l(x) - lP_{l-1}(x) \tag{A.19}$$

This is also satisfied by the irregular Legendre functions $Q_l(x)$. The $Q_l(x)$ can be calculated from $Q_0(x)$ using (A.19) (AS, 8.4.2)

$$Q_0(x) = \tfrac{1}{2}(\ln(1+x) - \ln(1-x)) \tag{A.20}$$

Travelling wave functions (Fuller & Moffa, 1977)

$$Q_l^{\pm}(x) = \tfrac{1}{2}(P_l(x) \mp i(2/\pi)Q_l(x)) \tag{A.21}$$

The following asymptotic forms for large $l$ are useful (Nussenzweig, 1965). If $\theta$ is not near zero or $\pi$ and if $l + \frac{1}{2} = \lambda$

$$P_l(\cos\theta) \sim \left[\frac{2}{\pi\lambda\sin\theta}\right]^{\frac{1}{2}} \left[\cos(\lambda\theta - \tfrac{1}{4}\pi) + \frac{\cot\theta}{8\lambda}\sin(\lambda\theta - \tfrac{1}{4}\pi) + \cdots\right] \tag{A.22}$$

$$Q_l^{\pm}(\cos\theta) \sim \frac{\exp\{\pm i(\lambda\theta - \frac{1}{4}\pi)\}}{(2\pi\lambda\sin\theta)^{\frac{1}{2}}}[1 \mp i(\cot\theta/8\lambda) + \cdots] \tag{A.23}$$

If $\theta$ is not too near to $\pi$ then the Szegö (1934) formula is useful

$$P_l(\cos\theta) \sim \left(\frac{\theta}{\sin\theta}\right)^{\frac{1}{2}} [J_0(\lambda\theta) + (\theta\cot\theta - 1)J_1(\lambda\theta)/(8\lambda\theta) + \cdots] \tag{A.24}$$

It is equivalent to (A.22) when $\theta \gg l^{-1}$ using (A.15) but is also accurate for $\theta \simeq 0$. Some of these formulae generalize to spherical harmonics. If $l^{-1} \ll \theta \ll \pi - l^{-1}$

$$Y_{l-m}(\theta, 0) \sim (-1)^{\frac{1}{2}(m-|m|)} \left(\frac{2l+1}{4\pi}\right)^{\frac{1}{2}} \left[\frac{(l-m)!(l+m)!}{l!^2}\right]^{\frac{1}{2}} \times \cos(\lambda\theta + \tfrac{1}{2}m\pi - \tfrac{1}{4}\pi) \tag{A.25}$$

This formula can be derived from (AS, 8.10.7). If $l \gg m^2$ it simplifies to

$$Y_{l-m}(\theta, 0) \sim (-1)^{\frac{1}{2}(m-|m|)} \left(\frac{2l+1}{4\pi}\right)^{\frac{1}{2}} \cos(\lambda\theta + \tfrac{1}{2}m\pi - \tfrac{1}{4}\pi) \tag{A.26}$$

Potgieter & Frahn (1967) give a generalization of (A.24) which is valid if $m^2(1/\theta - \cot\theta) \ll 2l + 1$

$$Y_{l-m}(\theta, 0) \sim \left(\frac{2l+1}{4\pi}\right)^{\frac{1}{2}} \left[\frac{(l-|m|)!}{(l+|m|)!}\right]^{\frac{1}{2}} (l + \tfrac{1}{2})^{|m|} \times \left(\frac{\theta}{\sin\theta}\right)^{\frac{1}{2}} J_m(\lambda\theta) \tag{A.27}$$

If $l > m^2$ a more approximate but simpler formula is (Austern & Blair, 1965)

$$Y_{l-m}(\theta, 0) \sim \left[\frac{2l+1}{4\pi} \cdot \frac{\theta}{\sin\theta}\right]^{\frac{1}{2}} J_m(\lambda\theta) \tag{A.28}$$

The Bessel function for $m < 0$ is given by the symmetry properties (A.13).

APPENDIX B

# The stationary phase and saddle point approximations

## B1 Gaussian integrals

The gaussian integral

$$I(\alpha) = \int_{-\infty}^{\infty} \exp\left(\tfrac{1}{2}i\alpha x^2\right)dx \tag{B.1}$$

converges if $\alpha$ has a positive imaginary part and has the value

$$I(\alpha) = (2\pi i/\alpha)^{\frac{1}{2}} = (2\pi/|\alpha|)^{\frac{1}{2}}e^{-i\chi}$$

$$\chi = \tfrac{1}{2}\arg\alpha - \tfrac{1}{4}\pi \quad \text{if } 0 < \arg\alpha < \pi \tag{B.2}$$

Equation (B.2) can also be used in the limiting case where $\alpha$ is real. Then

$$\chi = -\tfrac{1}{4}\pi \text{ if } \alpha > 0, \quad \chi = \tfrac{1}{4}\pi \text{ if } \alpha < 0 \tag{B.3}$$

Consider the multi-dimensional gaussian integral

$$I(A) = \int_{-\infty}^{\infty} \cdots \int_{-\infty}^{\infty} dx_1 \ldots dx_n \exp\left(\tfrac{1}{2}iA_{ij}x_i x_j\right) \tag{B.4}$$

where $A_{ij}$ is an $n \times n$ symmetric matrix and we use the convention of summing over repeated indices. If the matrix $A_{ij}$ is real the integral (B.4) can be evaluated by diagonalizing it. Then (B.4) reduces to a product of one-dimensional integrals like (B.1) and

$$I(A) = \prod_{i=1}^{n} (2\pi/|\alpha_i|)^{\frac{1}{2}} e^{-i\chi}$$

$$\chi = \tfrac{1}{2}m\pi - \tfrac{1}{4}n\pi \tag{B.5}$$

where $\alpha_i$ are the eigenvalues of $A_{ij}$ and $m$ is the number of negative eigenvalues. Equation (B.5) can also be written as

$$I(A) = [(2\pi)^n/|\det A|]^{\frac{1}{2}} e^{-i\chi} \tag{B.6}$$

and can be used when $A_{ij}$ is a complex symmetric matrix with a positive definite imaginary part. Then

$$\chi = \tfrac{1}{2}\arg(\det A) - \tfrac{1}{4}n\pi$$

## B2 Stationary phase approximation

The stationary phase method gives an approximate formula for the integral

$$I = \int_a^b g(z)\exp\left(\frac{i}{\hbar}\psi(z)\right)dz \tag{B.7}$$

where $g(z)$ and $\psi(z)$ are real functions of a real variable $z$. The idea is that in the limit, where $\hbar$ is small, the exponential in (B.7) is a rapidly oscillating function of $z$ except in the neighbourhood of points $z_\alpha$ where $\psi(z)$ is stationary

$$(d\psi/dz)_{z=z_\alpha} = \psi'(z_\alpha) = 0 \tag{B.8}$$

Then one can assume that the main contributions to the integral come from regions near these stationary points because then the contributions from nearby points will be in phase and will add constructively. In other regions contributions from neighbouring points will cancel because of the rapidly varying phase. If the stationary points are isolated from one another the contribution $I_\alpha$ of each $z_\alpha$ can be obtained by approximating the integrand by a gaussian function and replacing the limits $(a, b)$ by $(-\infty, \infty)$.

$$\begin{aligned} I_\alpha &\simeq g(z_\alpha)\int_{-\infty}^{\infty} \exp[(i/\hbar)(\psi(z_\alpha) + \tfrac{1}{2}(z - z_\alpha)^2\psi''(z_\alpha))]dz \\ &= g(z_\alpha)\left[\frac{2\pi i\hbar}{\psi''(z_\alpha)}\right]^{\frac{1}{2}} \exp[(i/\hbar)\psi(z_\alpha)] \end{aligned} \tag{B.9}$$

and the total integral is given by

$$I = \sum_\alpha I_\alpha \tag{B.10}$$

where the sum is taken over all the stationary points with $a < z_\alpha < b$. The width $\Delta z_\alpha$ of the gaussian in the integrand for $I_\alpha$ is given by

$$(\Delta z_\alpha)^2 = \hbar\psi''(z_\alpha)$$

and goes to zero as $\hbar \to 0$. The leading corrections to $I_\alpha$ can be obtained by multiplying it by an asymptotic series $S_\alpha$ in powers of $\hbar$. The first few terms in the series are

$$\begin{aligned} S_\alpha = 1 + (\hbar/i)[&\tfrac{1}{2}g''(z_\alpha)/(g(z_\alpha)\psi''(z_\alpha)) + \tfrac{1}{2}g'(z_\alpha)\psi'''(z_\alpha)/(g(z_\alpha)\psi''(z_\alpha)^2) \\ &+ (5/24)(\psi'''(z_\alpha))^2/(\psi''(z_\alpha))^3 + (1/8)\psi''''(z_\alpha)/(\psi''(z_\alpha))^2] + O(\hbar^2) \end{aligned} \tag{B.11}$$

The approximation (B.9) is accurate only if all the correction terms in (B.11) are small compared with unity. This condition is satisfied in the limit $\hbar \to 0$. There is an ambiguity in the sign of the factor in equation (B.9) because of the square root. To make the correct choice of sign explicit we write

$$I=\sum_\alpha g(z_\alpha)\left|\frac{2\pi\hbar}{\psi''(z_\alpha)}\right|^{\frac{1}{2}}\exp\left[(i/\hbar)\psi(z_\alpha)-i\chi_\alpha\right] \tag{B.12}$$

with

$$\chi_\alpha=-\tfrac{1}{4}\pi \quad \text{if } \psi''(z_\alpha)>0$$
$$\chi_\alpha=\tfrac{1}{4}\pi \quad \text{if } \psi''(z_\alpha)<0$$

## B3 Saddle point method

The purpose of the method is to give an approximate formula for the integral (B.7) where $g(z)$ and $\psi(z)$ can be complex. If the integrand is analytic in a certain domain the integration contour can be deformed within that domain. In order to get an approximate formula for $I$ it is advantageous to keep the contour in regions where the integrand is as small as possible. One can introduce a topography by taking $|\exp[(i/\hbar)\psi(z)]|$ as an altitude for each complex number $z$. The shape of this surface defines valleys and mountains. Curves with $\operatorname{Im}\psi(z)=\text{const}$ are contours of constant altitude. Curves with $\operatorname{Re}\psi(z)=\text{const}$ are orthogonal to these and define paths of steepest descent. The most favoured contours are those which stay in valleys, except for some passages from one valley to another over saddles given by

$$(d\psi/dz)_{z_\alpha}=\psi'(z_\alpha)=0 \tag{B.13}$$

The dominant contributions to the integral come from the neighbourhood of saddle points. Again approximating the integrand at each saddle point by a gaussian we obtain the same results (B.9) as for the stationary phase method. Note that not all solutions of (B.13) contribute. The sum in (B.9) is over all saddles crossed by the new integration contour. The topography of the integrand determines the sequence of saddles to be crossed so that there is a unique result for $I$.

If the result of the saddle point method is written in the form (B.12) the phase $\chi_\alpha$ is given by

$$\chi_\alpha=-\tfrac{1}{4}\pi+\tfrac{1}{2}\arg\psi''(z_\alpha) \tag{B.14}$$

In other words, the phase $\chi_\alpha$ is determined by the orientation of the steepest descent path at the saddle. Fig. B.1 shows some typical examples.

The stationary phase approximation and the saddle point method are closely related and in situations where both are applicable they give the same result. The saddle point method is more general. For example, in the case of rainbow scattering (section 4.4) there are two points of stationary phase in

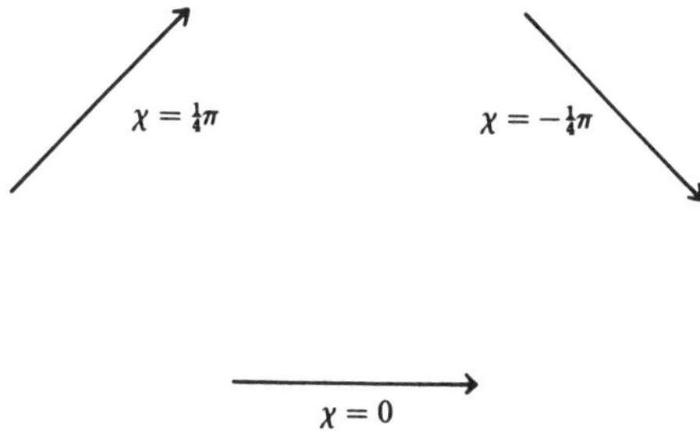


Fig. B.1. Relation between values of $\chi$ and the orientation of the steepest descent contour at a saddle point.

the rainbow integral when the scattering angle $\theta$ is less than the rainbow angle $\theta_R$ and both the stationary phase and saddle point approximations give the same result. As $\theta \to \theta_R$ the stationary points approach one another and both approximations fail. When $\theta > \theta_R$ the stationary points become complex and the stationary phase approximation cannot be used. The saddle point method is still applicable, only one of the two saddles contributes to the integral and the topography of the integrand determines which saddle is crossed.

The saddle point method can also be used for integrands which are not analytic everywhere but which have poles or branch points. Corrections to the saddle point method are given by equation (B.11). Both methods must be modified if there is a stationary point at or near one of the integration limits $a$ or $b$. Both the stationary phase and the saddle point method break down if two stationary points are close together, that is, when the spacing is comparable with the width $\Delta z_\alpha$. Then a uniform approximation can be used.

## B4 Stationary phase approximation in several dimensions

Consider an integral

$$I = \int_R g(z) \exp[(i/\hbar)\psi(z)] d^n z \tag{B.15}$$

where $g$ and $\psi$ are functions of $n$ real variables $\mathbf{z} = (z_1, \ldots, z_n)$ and $R$ is the domain of integration. In the limit where $\hbar \to 0$ the situation is similar to the one-dimensional case and the main contributions to $I$ come from stationary points $z_i^\alpha$ of the phase $\psi(z)$, that is

$$(\partial\psi/\partial z_i)_{z^\alpha} = \psi_i^\alpha = 0 \tag{B.16}$$

where $\psi_i^\alpha(z^\alpha)$ is the partial derivation of $\psi$ with respect to $z_i$ evaluated at the stationary point $\mathbf{z}^\alpha$. Making a gaussian approximation at each stationary point gives the result

$$I = \sum_\alpha g(\mathbf{z}^\alpha)[(2\pi\hbar)^n/|D_\alpha|]^{\frac{1}{2}} \exp[(i/\hbar)\psi(\mathbf{z}^\alpha) - \chi_\alpha] \tag{B.17}$$

In equation (B.17) $D_\alpha = \det \psi_{ij}^\alpha$ is the determinant of the matrix of second derivatives of $\psi$ at the stationary point $\mathbf{z}^\alpha$. The phase $\chi_\alpha$ is given by

$$\chi_\alpha = \tfrac{1}{2}m\pi - \tfrac{1}{4}n\pi \tag{B.18}$$

where $m$ is the number of negative eigenvalues of the matrix $\psi_{ij}^\alpha$. The summation in (B.17) is over all stationary points inside the region $R$ of integration. If two stationary points coincide then $D_\alpha = 0$ and the method breaks down. If two stationary points are close together or if one is near the boundary of the region $R$ the corrections are large.

In the case of a one-dimensional integral the saddle point method is a generalization of the stationary phase approximation. There is a similar generalization for multi-dimensional integrals of the form (B.15) where $g(z)$ and $\psi(z)$ are analytic functions of several variables $z_1, \ldots, z_n$. The result is still given by equation (B.18) where $\mathbf{z}^\alpha$ are the solutions of the stationary conditions (B.16) but there are differences. It is more difficult to decide which stationary points $\mathbf{z}^\alpha$ should be retained in the summation in equation (B.17). This question can be answered in the one-dimensional case by studying the topography of the function $|\exp((i/\hbar)\psi(z))|$. This determines the sequence of saddles to be crossed and leads to a unique result for the integral. It is almost impossible to follow this procedure with a function of several complex variables and it may be necessary to rely on physical arguments to pick out the stationary points that contribute.

APPENDIX C

# Uniform approximation for integrals

## C1 One-dimensional integrals

We want to calculate the integral

$$I = \int_a^b g(z)\exp((i/\hbar)\psi(z))dz \tag{C.1}$$

The function $\psi$ has stationary points $z_\alpha$ which satisfy

$$(d\psi/dz)_{z_\alpha} = \psi'(z_\alpha) = 0 \tag{C.2}$$

The integral can be evaluated approximately either by the stationary phase or the saddle point method to give (B.9)

$$\begin{aligned} I &= \sum_\alpha A_\alpha \exp((i/\hbar)\psi(z_\alpha)) \\ A_\alpha &= g(z_\alpha)[2\pi i\hbar/\psi''(z_\alpha)]^{\frac{1}{2}} \end{aligned} \tag{C.3}$$

where the sum is over all contributing stationary points. If two or more stationary points $z_\alpha$ coincide (C.3) diverges because $\psi''(z_\alpha)=0$ when stationary points come together and both methods fail. When stationary points are close together the main contribution to $I$ still comes from near these points but the primitive approximation (B.9) is inaccurate. In these circumstances a uniform approximation can be used to obtain a more accurate result. Such uniform approximations were developed by Chester *et al.* (1957) and Maslov (1972) and have been applied to scattering problems by Berry (1966), Miller (1968*a*,*b*) da Silveira (1973), Knoll & Schaeffer (1976, 1977) and others.

The idea of the uniform approximation is to change the integration variable $z$ in (C.1) to a new variable $y$ by a mapping

$$\psi(z) = \phi(\beta, y) \tag{C.4}$$

where $\phi(\beta,y)$ is a simple function of $y$ which has the same structure of stationary points as $\psi(z)$. This function also depends on one or several parameters $\beta$. The mapping (C.4) is one-to-one provided

$$dz/dy = (d\phi/dy)/(d\psi/dz) \neq 0 \tag{C.5}$$

and is finite. This requires that the stationary points $z_\alpha$ of $\psi$ and $y_\alpha$ of $\phi$ map onto each other

$$\psi(z_\alpha) = \phi(\beta, y_\alpha) \tag{C.6}$$

This relation fixes the parameters $\beta$. The transformed integral is now

$$I = \int_{\bar{a}}^{\bar{b}} \exp((i/\hbar)\phi(y))g(z(y))(dz/dy)dy \tag{C.7}$$

where $\bar{a}$ and $\bar{b}$ are the transformed end points. Equation (C.7) is exact but now the whole complexity of the integral is in the product $g(z)(dz/dy)$. The approximation is to choose a function

$$h(y) \simeq g(z(y))(dz/dy) \tag{C.8}$$

Equation (C.8) is required to hold exactly at the stationary points and to be a good approximation for other values of $z$. If we also require that $\phi(y)$ and $h(y)$ are simple enough for the integral

$$J(\beta) = \int_{\bar{a}}^{\bar{b}} h(y) \exp((i/\hbar)\phi(\beta, y))dy \tag{C.9}$$

to be calculated analytically we have a useful approximation

$$I \simeq J(\beta) \tag{C.10}$$

for the integral (C.1). From equation (C.4)

$$\frac{d^2\phi}{dy^2} = \frac{d^2\psi}{dz^2}\left(\frac{dz}{dy}\right)^2 + \frac{d\psi}{dz}\frac{d^2z}{dy^2} \tag{C.11}$$

The second term in (C.11) is zero at the stationary points $z_\alpha$ and at those points

$$h(y_\alpha) = g(z_\alpha)[(d^2\phi/dy^2)/(d^2\psi/dz^2)]^{\frac{1}{2}} \tag{C.12}$$

Equations (C.6) and (C.12) which fix the parameters in $\phi(y)$ and $h(y)$ do not depend on $\hbar$. They also ensure that the stationary phase approximations to $I$ and $J(\beta)$ are identical. Hence $I$ and $J$ are asymptotically equal as $\hbar \to 0$. $J$ will be a good approximation to $I$ even when (C.3) is not, provided that equation (C.8) is reasonably accurate between stationary points.

It may not be possible to find a simple function $\phi$ which maps all the stationary points of $\psi$. Then $\phi$ should be chosen so that (C.6) holds for the physically important stationary points. In the next sections we give examples of some mapping functions which yield useful uniform approximations.

## C2 The uniform Airy approximation

This form is useful when the integral (C.1) has two stationary points $z_1$ and $z_2$ and $\psi(z)$ has an almost cubic structure. The mapping function $\phi$ is taken to be

$$\phi(y)/\hbar = \eta - \beta y - \tfrac{1}{3}y^3 \tag{C.13}$$

with two parameters $\eta$ and $\beta$, and the amplitude

$$h(y) = (g_1 + g_2 y)/(2\pi) \tag{C.14}$$

The mapping function $\phi$ has stationary points $y_1 = \sqrt{\beta}$ and $y_2 = -\sqrt{\beta}$. Equation (C.6) fixed $\eta$ and $\beta$

$$\eta = (\psi(z_1) + \psi(z_2))/2\hbar, \quad \beta = [(3/4\hbar)(\psi(z_2) - \psi(z_1))]^{\frac{2}{3}} \tag{C.15}$$

and (C.12) determines the values of $q_1$ and $g_2$

$$\begin{aligned} g_1 + g_2\sqrt{\beta} &= 2\pi^{\frac{1}{2}}\beta^{\frac{1}{4}}A_1 e^{-\frac{1}{4}i\pi} \\ g_1 - g_2\sqrt{\beta} &= 2\pi^{\frac{1}{2}}\beta^{\frac{1}{4}}A_2 e^{\frac{1}{4}i\pi\pi} \end{aligned} \tag{C.16}$$

where $A_1$ and $A_2$ are given in (C.3). Evaluating the integral (C.9) in terms of Airy functions leads to

$$I \simeq (g_1 Ai(-\beta) - ig_z A_i'(-\beta))e^{i\eta} \tag{C.17}$$

## C3 The uniform Bessel approximation

The Bessel function approximation is useful if (C.1) has two saddle points and the variable $z$ is periodic, that is, if $a = 0$, $b = 2\pi$, $g(z + 2\pi) = g(z)$ and $\psi(z + 2\pi) = \psi(z)$. This time we choose

$$\phi(y) = \eta + \beta\cos y \tag{C.18}$$

$$h(y) = (g_1 + ig_2\cos y)i^{-M}e^{iMy}/(2\pi) \tag{C.19}$$

and the integral for $J(\beta)$ can be expressed in terms of Bessel functions

$$J = e^{i\eta}(g_1 J_M(\beta) + g_2 J'_M(\beta)) \tag{C.20}$$

The constants $\eta$ and $\beta$ can be found from (C.6)

$$\begin{aligned} \eta &= (\psi(z_1) + \psi(z_2))/2\hbar \\ \beta &= (\psi(z_1) - \psi(z_2))/2\hbar \end{aligned} \tag{C.21}$$

while $g_1$ and $g_2$ are fixed by (C.12) in terms of $A_1$ and $A_2$

$$\begin{aligned} g_1 + ig_2 &= \sqrt{(2\pi\beta)}A_1\exp(-(M + \tfrac{1}{2})i\pi/2) \\ g_1 - ig_2 &= \sqrt{(2\pi\beta)}A_2\exp((M + \tfrac{1}{2})i\pi/2) \end{aligned} \tag{C.22}$$

The integer $M$ is fixed by the requirement that (C.8) should be a reasonable approximation for all $z$.

## C4 Multi-dimensional integrals

The uniform approximation can be extended to apply to multi-dimensional integrals. In this appendix we discuss the generalization at the most primitive level possible and refer to Levit & Smilansky (1977*b*) for a more complete account. The extension is used in chapter 7 for the evaluation of certain path integrals. Consider an n–N-dimensional integral

$$I = \int_R g(z) \exp((i/\hbar)\psi(z)) d^N z \tag{C.23}$$

where $g(z)$ and $\psi(z)$ are real functions of $N$ real variables $(z_1, \ldots, z_n) = \mathbf{z}$. The idea is to consider an integral

$$J = \int_R h(y) \exp((i/\hbar)\phi(\beta, \mathbf{y})) d^n y \tag{C.24}$$

which is simpler than (C.23) but which has the same structure of stationary points. In (C.24) $h(\mathbf{y})$ and $\phi(\beta,\mathbf{y})$ are functions of $n$ real variables $\mathbf{y} = (y_1, \ldots, y_n)$ and $\beta = (\beta_1, \ldots, \beta_m)$ are $m$ parameters.

The integral $J(\beta)$ is a uniform approximation to $I$ if there is a correspondence between the stationary points of $\psi$ and those of $\phi$. We assume that the parameters $\beta$ can be chosen so that both $\psi$ and $\phi$ and $p$ stationary points in the respective regions of integration $R$ and $\bar{R}$ at $\mathbf{z}^\alpha$ and $\mathbf{y}^\alpha$, $\alpha = 1, \ldots, p$ which are determined by the conditions

$$(\partial\psi/\partial z_i)_{\mathbf{z}^\alpha} = \psi_i^\alpha = 0, \quad i = 1, \ldots, N, \quad \alpha = 1, \ldots, p$$
$$(\partial\phi/\partial y_j)\mathbf{y}^\alpha = \phi_j^\alpha = 0, \quad j = 1, \ldots, n, \quad \alpha = 1, \ldots, p \tag{C.25}$$

At each stationary point the following relations should hold

$$\psi(z^\alpha) = \phi(\beta, y^\alpha) \tag{C.26}$$

$$g(z^\alpha)\left[\frac{(2\pi i\hbar)^N}{\det \psi_{ij}^\alpha}\right]^{\frac{1}{2}} = h(y^\alpha)\left[\frac{(2\pi i\hbar)^n}{\det \phi_{ij}^\alpha}\right]^{\frac{1}{2}} \tag{C.27}$$

In (C.27) $\psi_{ij}^\alpha$ and $\phi_{ij}^\alpha$ are the matrices of second partial derivatives of $\psi$ and $\phi$ evaluated at the stationary point $\alpha$. If (C.26) and (C.27) hold for all the stationary solutions of (C.25) in the regions $R$ and $\bar{R}$ then the evaluations of the integrals (C.23) and (C.24) by the stationary phase method give identical results. Thus, when the stationary phase method gives a good approximation, $J$ is a good approximation to $I$. The hope of the uniform method is that $J$ continues to be a good approximation to $I$ even when the stationary phase method fails. If (C.26) and (C.27) also hold for some complex stationary points then the uniform approximation takes their contribution into account to some extent.

The uniform approximation has a number of very nice features. If $N \gg n$ it gives an approximation to an integral in many dimensions by one in a much smaller number of dimensions. This is especially useful in the context of path integrals. The comparison integral $J$ over the real variables $y_1, \ldots, y_n$ can be evaluated either analytically or by using standard numerical methods. The uniform approximation uses the same input as the stationary phase approximation and gives finite answers in situations where the stationary phase method fails. When applying the saddle point method to multi-dimensional integrals it may be difficult to decide which stationary points to include in the summation in equation (B.17). The uniform approximation avoids this difficulty. If all the stationary points, which could be relevant, are included in the mapping (C.26) and (C.27) then evaluating the integral $J$ automatically gives the correct contribution from each one.

There are, however, some major pitfalls. There may be several choices of mapping functions $\phi$ and $h$ which give identical results for $J$ in the stationary phase limit, but which differ away from that limit. How can one decide which is the correct choice? This difficulty is present even in the one-dimensional uniform Bessel approximation. Any choice of the integer $M$ in (C.19) gives the same asymptotic formula for $J$ in (C.20) when $\beta$ is large, but the results are very different for small $\beta$. In the one-dimensional case $M$ is fixed by the requirement that (C.8) should be a good approximation for all $z$. There is no analogous requirement in the multi-dimensional case, because the integrals (C.23) and (C.24) have a different number of dimensions.

There are some guidelines for choosing the mapping functions when uniform approximations are used for calculating wave fields in the short wave length limit. The important ideas have been reviewed by Berry (1976). In such applications the integral $I$ in (C.23) is a wave amplitude and depends on some parameters specifying the position in the wave field. The stationary phase evaluation of $I$ corresponds to a description of the wave field by a family of trajectories or rays. If the family exhibits a focusing property the stationary phase formula (B.17) for $I$ becomes infinite on the caustics in the focal region, but the physical amplitude of the wave is finite. A uniform approximation goes over to the stationary phase result, away from the caustics and gives an accurate asymptotic approximation in the focal region.

The form of the integral $J$ of the uniform approximation (C.24) depends on the structure of the caustics. Berry (1966, 1969) and Berry & Mount (1972) have devised uniform approximations for various special types of caustic and more recently a general theory has been developed. The theory of Thom (1975) enables certain caustics – those which are structurally stable – to be classified according to their topology. Thom's theory is sometimes called 'catastrophy theory' and the caustics are referred to as catastrophies. The derivation of uniform asymptotic approximations to wave fields in the presence of these caustics has been put on a rigorous basis by Maslov (1972), Duistermaat (1974) and others. The conclusion which follows from all these investigations is that $J$ is a valid uniform approximation to $I$ only if $I$ and $J$ have the same structure of caustics and if the stationary phase evaluations agree outside the caustic region.

APPENDIX D

# Uniform approximations for Schrödinger's equation

This method for obtaining approximate solutions of second order differential equations was set out by Miller & Good (1953) and by Dingle (1956). In this appendix we follow the discussion given in the review article by Berry & Mount (1972). Another kind of uniform approximation for evaluating integrals, and which is related to the saddle point approximation, is presented in appendix C.

We write Schrödinger's equation as

$$(d^2\psi/dx^2) + \chi(x)\psi = 0 \tag{D.1}$$

where

$$\begin{aligned} \chi(x) &= k^2(x) = (2m/\hbar^2)(E - V(x)); \quad && E > V(x) \\ &= -\gamma^2(x); && E < V(x) \end{aligned}$$

and consider a comparison equation for a function

$$(d^2\phi/d\xi^2) + \Gamma(\xi)\phi = 0 \tag{D.2}$$

The idea of the uniform approximation is that $\Gamma(\xi)$ should be chosen to be qualitatively similar to $\chi(x)$ but simpler so that the solutions of (D.2) are known functions. Because $\Gamma(\xi)$ is similar to $\chi(x)$, the function $\phi(\xi)$ should be similar to $\psi(x)$ and can be changed into it by stretching or contracting it and by changing its amplitude. Following Berry & Mount (1972) and Jeffries & Jeffries (1956, §17.13) we put

$$\psi(x) = (d\xi/dx)^{-\frac{1}{2}}\phi(\xi) \tag{D.3}$$

If $\psi(x)$ satisfies (D.1) then

$$(\xi')^2(d^2\phi/d\xi^2) + \chi\phi = [\tfrac{1}{2}(\xi'''/\xi') - \tfrac{3}{4}(\xi''/\xi')^2]\phi \tag{D.4}$$

Now choose

$$\xi' = (d\xi/dx) = (\chi(x)/\Gamma(\xi))^{\frac{1}{2}} \tag{D.5}$$

or

$$\int_{\xi_0}^{\xi} \sqrt{\Gamma(\xi)}d\xi = \int_{x_0}^{x} \sqrt{\chi(x)}dx \tag{D.6}$$

Combining (D.4) and (D.5) gives

$$(d^2\phi/d\xi^2) + \Gamma(\xi)\phi = g(\xi)\phi$$

where

$$g(\xi) = (\Gamma(\xi)/\chi(x))[\tfrac{1}{2}(\xi'''/\xi') - \tfrac{3}{4}(\xi''/\xi')^2]$$
$$= \Gamma(\xi)\varepsilon(\xi)$$

Hence $\psi(x)$ given by (D.3) is a good approximation to the solution of the Schrödinger equation provided $\phi$ satisfies (D.2) and $g$ can be neglected. This is so if

$$|\varepsilon(\xi)| \ll 1. \tag{D.7}$$

If the functions $\Gamma(\xi)$ and $\chi(x)$ have zeros then the lower end points of the integral (D.6) should be corresponding zeros of $\Gamma$ and $\chi$. Also the transformation (D.6) should map zeros of $\chi(x)$ onto zeros of $\Gamma(\xi)$. When the mapping is made in this way then $\xi'$ can be finite and non-zero and $\varepsilon(\xi)$ can be small even at turning points when $\chi = 0$.

We now discuss some special cases:

(i) Choose $\Gamma(\xi) = 1$, $\phi(\xi) = \exp(\pm i\xi)$

Equations (D.5) and (D.6) give

$$\xi' = \sqrt{\chi(x)} = k(x),\ \xi = \int k(x)dx \tag{D.8}$$

Substituting (D.8) into (D.3) gives the WKB approximation.

(ii) Suppose $\chi(x) = 0$ at $x = a$, $\chi(x) > 0$ for $x > a$ and $\chi(x) < 0$ for $x < a$ so that $a$ is a left turning point. Choose $\Gamma(\xi) = \xi$ and map $x = a$ onto $\xi = 0$. Equations (D.5) and (D.6) give

$$\xi' = (\chi(x)/\xi)^{\frac{1}{2}} > 0, \quad (2/3)\xi^{\frac{3}{2}} = \int_a^x k(x)dx$$

The solutions of equation (D.2) are Airy functions $\mathrm{Ai}(-\xi)$ and $\mathrm{Bi}(-\xi)$. When $|\xi|$ is large the Airy function $\mathrm{Ai}(-\xi)$ can be replaced by its asymptotic forms (A.5) or (A.6) then the approximate solution (D.3) of the Schrödinger equation goes over to one of the WKB solutions. Thus the uniform approximation provides a connection formula for the two WKB solutions on opposite sides of the turning point. The result is the same as in (3.20).

(iii) Choose $\Gamma = \xi^2/4 - \beta$. This corresponds to a parabolic potential barrier and the comparison equation can be used to find approximate solutions for barrier penetration problems. Fig. D.1 shows a typical potential barrier and the comparison function $\Gamma(\xi)$. The allowed regions $x < b$ and $x > c$ should be mapped onto the allowed regions $|\xi| > 2\sqrt{\beta}$ of the comparison problem while the forbidden region $b < x < c$ should be mapped onto $-2\sqrt{\beta} < \xi < 2\sqrt{\beta}$. If the turning points $b$ and $c$ are to be

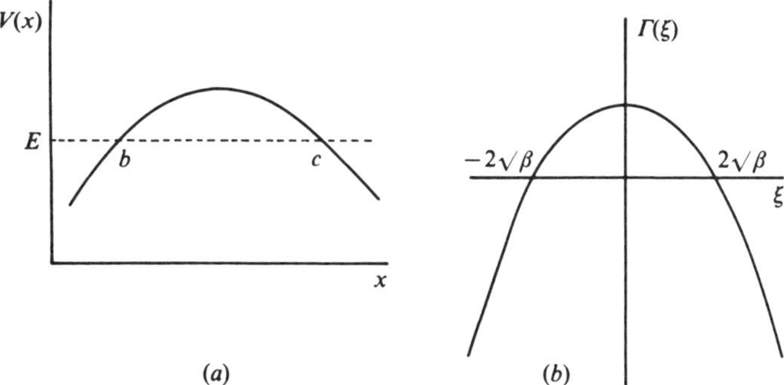


Fig. D.1. (a) A typical potential barrier $V(x)$, (b) The parabolic comparison function $\Gamma(\xi)$.

mapped onto $\pm 2\sqrt{\beta} = \pm\alpha$ then the mapping condition (D.6) requires that

$$\int_b^c \gamma(x)dx = \int_b^c \sqrt{(-\chi(x))}dx = \int_{-\alpha}^{\alpha} \sqrt{(\beta - \tfrac{1}{4}\xi^2)}d\xi = \pi\beta \qquad \text{(D.9)}$$

Thus $\beta$ has to be chosen to satisfy (D.9)

The two independent solutions of the comparison equation are Whittaker functions. Discussion of the barrier penetration problem will be continued in appendix E.

APPENDIX E

# The potential barrier

The penetration of a particle through a potential barrier can be calculated using the matching formulae given in section 3.3 or by the parabolic uniform approximation of appendix D. A typical barrier is shown in fig. E.1. The incident energy is below the top of the barrier and there are turning points at $x=b$ and $x=c$. The region $b<x<c$ is classically inaccessible. We suppose that a wave with unit amplitude is incident on the barrier from the left hand side and that there is a reflected wave with amplitude $R$. On the right-hand side of the barrier there is a transmitted wave with amplitude $T$. Thus

$$\psi(x)=k^{-\frac{1}{2}}\left[\exp\left(i\int_b^x k(x)dx\right)+R\exp\left(-i\int_b^x k(x)dx\right)\right]; \quad x<b \quad \text{(E.1)}$$

$$=k^{-\frac{1}{2}}T\exp\left(i\int_c^x k(x)dx\right); \quad x>c \quad \text{(E.2)}$$

In the simple approach (see Landau & Lifshitz, 1958) it is assumed that the solution in the region $b<x<c$ is a superposition of an exponentially decreasing and increasing wave

$$\psi(x)=\gamma^{-\frac{1}{2}}\left[B_-\exp\left(-\int_b^x \gamma(x)dx\right)+B_+\exp\left(\int_b^x \gamma(x)dx\right)\right] \quad \text{(E.3)}$$

These two components have comparable magnitudes near $x=c$ but if the barrier is thick in the sense that

$$Q=\int_b^c \gamma(x)dx \gg 1 \quad \text{(E.4)}$$

then the increasing component proportional to $B_+$ is negligible compared with the other component near $x=b$. Their ratio is approximately $\exp(-2Q)$. Hence, when matching the solutions to the left and right of $x=b$ it is only necessary to consider the component $B_-$. Then the connection formula (3.19) or better (3.26) gives

$$B_-=\exp(-i\pi/4), \quad R=\exp(-i\pi/2) \quad \text{(E.5)}$$

To find the transmission amplitude we use the conncection formula (3.23) at

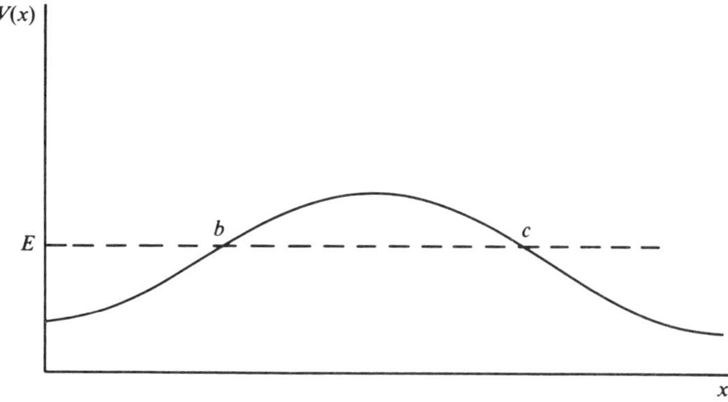


Fig. E.1. A potential barrier $V(x)$ with classical turning points ($V(x)=E$) at $x=b$ and $x=c$.

$x=c$. Now the component proportional to $B_-$ is exponentially increasing away from $C$ and $D$ is determined as

$$D = B_- \exp(-Q)$$

because

$$-\int_b^x \gamma(x)dx = \int_x^c \gamma(x)dx - Q$$

The coefficient $E$ has to be chosen so that the oscillating solution corresponds to an outgoing wave, i.e. $E=D$. Combining these results together gives

$$T = D\exp(i\pi/4) = \exp(-Q) \tag{E.6}$$

The interesting point about the derivation is that the second coefficient $B_+$ in the wave under the barrier is not required. This is the consequence of the thick barrier condition (E.4).

If the incident energy increases near to the barrier top then $Q$ becomes smaller and the assumptions made in deriving (E.5) and (E.6) are no longer justified. But in the same situation the whole WKB approach becomes inaccurate because the validity condition (3.6) is not satisfied. It is necessary to use a completely different method and the uniform approximation is suitable for the purpose.

We use the parabolic comparison function $\Gamma = \xi^2/4 - \beta$ and the known solutions of the comparison equation (D.2) are Whittaker functions. The general asymptotic form of a solution of the comparison equation on the right-hand side of the parabolic barrier is

$$\phi(\xi) \sim (k_0(\xi))^{-\frac{1}{2}}[C_+ \exp(iW_0(\xi,\alpha)) + C_- \exp(-iW_0(\xi,\alpha))] \tag{E.7}$$

and on the left-hand side it is

$$\phi(\xi) \sim (k_0(\xi))^{-\frac{1}{2}}[B_+ \exp(iW_0(\xi, -\alpha)) + B_- \exp(-iW_0(\xi, -\alpha))] \quad \text{(E.8)}$$

In these equations $\alpha = 2\sqrt{\beta}$ and $\xi = \pm\alpha$ are the turning points. The action integrals are

$$W_0(\xi, \alpha) = \int_\alpha^\xi k_0(\xi) d\xi \quad \text{for } \xi > \alpha$$

$$W_0(\xi, -\alpha) = \int_{-\alpha}^\xi k_0(\xi) d\xi \text{ for } \xi < -\alpha$$

$$k_0(\xi) = (\xi^2/4 - \beta)^{\frac{1}{2}}$$

A study of the known asymptotic forms of the Whittaker functions leads to a matrix relation between the coefficients $C_\pm$ and $B_\pm$ (see Connor, 1973)

$$\begin{pmatrix} C_+ \\ C_- \end{pmatrix} = \begin{pmatrix} f_{11} & -if_{12} \\ if_{21} & f_{22} \end{pmatrix} \begin{pmatrix} B_+ \\ B_- \end{pmatrix} = F \begin{pmatrix} B_+ \\ B_- \end{pmatrix} \quad \text{(E.9)}$$

where

$$f_{11} = e^{\pi\beta}\bar{N}(i\beta), \quad f_{22} = e^{\pi\beta}N(i\beta) \quad \text{(E.10)}$$

$$f_{12} = f_{21} = e^{\pi\beta} \quad \text{(E.11)}$$

$$N(i\beta) = \frac{(2\pi)^{\frac{1}{2}} \exp(-\pi\beta/2 + i(\beta \ln \beta - \beta))}{\Gamma(\frac{1}{2} + i\beta)} \quad \text{(E.12)}$$

$$\bar{N}(i\beta)N(i\beta) = 1 + \exp(-2\pi\beta) \quad \text{(E.13)}$$

Using Stirling's formula for the gamma function in (E.12) one obtains an asymptotic series in powers of $\beta^{-1}$

$$N(i\beta) \sim (1 - i/(12\beta) + \cdots)$$

The matrix $F$ has some general properties which are satisfied by the special forms (E.10). The first of these is

$$\det F = 1 \quad \text{(E.14)}$$

This can be proved by considering two independent solutions of the comparison equation with coefficients $C, B$ and $C', B'$. The coefficients are connected by the matrix relation

$$\begin{pmatrix} C_+ & C'_+ \\ C_- & C'_- \end{pmatrix} = F \begin{pmatrix} B_+ & B'_+ \\ B_- & B'_- \end{pmatrix}$$

The Wronskian of two independent solutions of the comparison equation is independent of $\xi$ and, in particular, has the same value on opposite sides of the barrier. This requires

$$C_+C'_- - C_-C'_+ = B_+B'_- - B_-B'_+$$

which in turns leads to (E.14). A second property of $F$ is that $f_{12} = f_{21}$. This is a consequence of the symmetry of the barrier. Because of these properties only two of the four elements of $F$ are independent. If $\beta$ is real then $f_{12}$ and $f_{21}$ are real while $f_{11} = f_{22}^*$. In this case $\bar{N}(i\beta) = N(i\beta)^*$ and

$$|f_{11}|^2 = |f_{22}|^2 = 1 + \exp(2\pi\beta) \tag{E.15}$$

The connection formula for waves on opposite sides of the potential barrier in fig. E.1 can now be found using the uniform approximation. Because of (D.6) it is only necessary to make the replacements

$$k_0(\xi) \to k(x)$$

$$W_0(\xi, \alpha) \to W(x, c) = \int_c^x k(x)dx \tag{E.16}$$

$$W_0(\xi, -\alpha) \to W(x, b) = \int_b^x k(x)dx \tag{E.17}$$

in (E.7) and (E.8). Then the connection formulae are given by (E.9) with $\beta$ given by

$$\pi\beta = \int_b^c \gamma(x)dx = Q \tag{E.18}$$

Transmission and reflection coefficients can be found by making a special choice of the coefficients in (E.9) and (E.8), namely $B_+ = 1$ and $C_- = 0$. Then $B_- = R$ and $C_+ = T$. Substituting into (E.9)

$$R = -if_{12}/f_{22}, \quad T = 1/f_{22} \tag{E.20}$$

In the case of a real barrier (E.10) and (E.20) give

$$R^2 = e^{2\pi\beta}/(1 + e^{2\pi\beta}), \quad T^2 = 1/(1 + e^{2\pi\beta}) \tag{E.21}$$

These formulae reduce to (E.5) and (E.6) if the incident energy is well below the top of the barrier, but they are also valid at the barrier and can be used to calculate reflection when the incident energy is above the barrier (Kemble, 1935).

APPENDIX F

# Fresnel integrals

The purpose of this appendix is to give some properties of Fresnel integrals and to derive the formula (5.15) for $J(\chi)$.

The Fresnel integrals $C(z)$ and $S(z)$ are defined as

$$C(z) = \int_0^z \cos\left(\tfrac{1}{2}\pi t^2\right) dt, \quad S(z) = \int_0^z \sin\left(\tfrac{1}{2}\pi t^2\right) dt$$

and the auxiliary functions $f(z)$ and $g(z)$ are related to them by

$$C(z) + iS(z) = \tfrac{1}{2}(1+i) - (g(z) + if(z)\exp\left(\tfrac{1}{2}\pi i z^2\right) \tag{F.1}$$

They are tabulated and have simple rational approximations for $z > 0$ (Abramowitz & Stegun 1965, 7.3)

$$f(z) \simeq \frac{1 + 0.926\,z}{2 + 1.792\,z + 3.104\,z^2} \tag{F.2}$$

$$g(z) \simeq \frac{1}{2 + 4.142\,z + 3.492\,z^2 + 6.670\,z^3}$$

which have an error $\varepsilon(z) < 10^{-3}$ for $0 < z < \infty$.

In order to derive formula (5.15) for $J$ we first introduce the integrals ($B = |\Theta'_{\rm g}|$)

$$G_>(\chi) = \int_{-\infty}^{0} d\mu \exp i\left(\chi\mu - \tfrac{1}{2}B\mu^2\right); \quad \chi > 0 \tag{F.3}$$

$$G_<(\chi) = \int_{0}^{\infty} d\mu \exp i\left(\chi\mu - \tfrac{1}{2}B\mu^2\right); \quad \chi < 0$$

These integrals can be evaluated by means of the substitutions

$$\mu = \chi/B + (\pi/B)^{\frac{1}{2}} z; \quad w = \chi/(\pi B)^{\frac{1}{2}}$$

$$G_<(-\chi) = G_>(\chi) = (\pi/B)^{\frac{1}{2}}(g(w) - if(w)) \simeq -i/\chi \quad \text{for } w \gg 1 \tag{F.4}$$

First we consider the case $\chi > 0$ and define

$$G(\chi) = \int_{-\infty}^{\infty} (1 - S_{\rm n}(\lambda_{\rm g} + \mu)) \exp i\left(\chi\mu - \tfrac{1}{2}B\mu^2\right) d\mu \tag{F.5}$$

$$= \int_{-\infty}^{\infty} d\mu (H(\mu) - S_{\rm n}(\lambda_{\rm g} + \mu)) \exp i\left(\chi\mu - \tfrac{1}{2}B\mu^2\right) + G_>(\chi) \tag{F.6}$$

where

$$H(\mu)=0 \quad \text{for } \mu<0$$
$$=1 \quad \text{for } \mu>0$$

The next step is to neglect the phase $\frac{1}{2}B\mu^2$ in the first integral in equation (F.6). This is a good approximation if $\chi \lesssim \pi/\Delta$ because the integrand is large only over the range $-\Delta \lesssim \mu \lesssim \Delta$ and the phase $\frac{1}{2}\mu^2 B$ is small over this range because of the condition (5.4) ($\frac{1}{2}\mu^2 B \ll 1$). With this approximation the integral can be evaluated with the result

$$\begin{aligned} G(\chi) &\simeq -(i/\chi)(F(\chi)-1)+G_>(\chi) \\ &= F(\chi)G_>(\chi)-(F(\chi)-1)((i/\chi)+G_>(\chi)) \end{aligned} \tag{F.7}$$

The second term in (F.7) is finite at $\chi=0$, because $F(0)=1$, and $(F(\chi)-1)$ is small over the whole range $0<\chi<\pi/\Delta$ if $\lambda_g$ is suitably chosen. Hence

$$G(\chi) \simeq F(\chi)G_>(\chi) \tag{F.8}$$

and

$$J(\chi) \simeq (B/2\pi)^{\frac{1}{2}} e^{\frac{1}{4}i\pi} F(\chi)G_>(\chi)$$

Using (F.4) for $G_>(\chi)$ we obtain (5.15). When $\chi$ is large ($\chi \gg \pi/\Delta$) the above argument cannot be used because of the discontinuity at $\mu=0$ in the integrand of the integral in (F.6). But then it can be replaced by a simpler argument. If $\chi>\pi\Delta$ the term $\frac{1}{2}B\mu^2$ in the integrand of (F.5) can be neglected and the integral evaluated to give

$$G(\chi) \simeq -(i/\chi)F(\chi) \tag{F.9}$$

But if $\chi>\pi\Delta$ then $w>(\pi/B\Delta^2)^{\frac{1}{2}} \gg 1$ and $G_>(\chi) \simeq -(i/\chi)$ so that (F.9) is equivalent to (F.8) and again yields the result (5.15).

The relation (5.15) for $\chi<0$ can be derived by a similar argument. Finally, the value of $G(0)$ is needed for the discussion in section 5.4. From equation (F.7)

$$G(0)=(\pi/2B)^{\frac{1}{2}} e^{-\frac{1}{4}i\pi} - iF'(0) \tag{F.10}$$

# References

Abramowitz, M. & Stegun, I. A. (1965). *Handbook of Mathematical Functions* New York: Dover Publ. Inc.

Albinski, J. & Michel, F. (1982). *Phys. Rev.*, **C25**, 213.

Alder, K. & Winther, A. (1966). *Coulomb Excitation*. New York: Academic Press.

Anni, R. & Taffara, L. (1974). *Nuovo Cimento*, **A29**, 595.

Anni, R. & Taffara, L. (1976). *Nuovo Cimento*, **A31**, 321.

Anni, R., Renna, L. & Taffara, L. (1978). *Nuovo Cimento*, **A45**, 123.

Anni, R., Renna, L. & Taffara, L. (1980). *Nuovo Cimento*, **A59**, 38.

Anni, R. & Renna, L. (1981*a*). *Lett. Nuovo Cimento* **30**, 229.

Anni, R. & Renna, L. (1981*b*). *Nuovo Cimento*, **A65**, 331.

Aoki, K. & Horiuchi, H. (1982). *Prog. Theoret. Phys.*, **68**, 1658.

Aoki, K. & Horiuchi, H. (1983). *Prog. Theoret. Phys.*, **69**, 1154.

Austern, N. (1965). *Phys. Rev.*, **137B**, 752.

Austern, N. & Blair, J. S. (1965). *Ann. Phys.*, **83**, 15.

Avishai, Y. & Knoll, J. (1976). *Z.* Phys., **A279**, 415.

Ball, J. B., Fulmer, C.B., Gross, E.E., Halbert, M. L., Hensley, D.C., Ludemann, C. A., Saltmarsh, M. J. & Satchler, G. R. (1975). *Nucl. Phys.*, **A252**, 208.

Bass, R. (1980). *Nuclear Reactions with Heavy-Ions*. Berlin: Springer-Verlag.

Bassani, G., Saunier, N., Traore, B. M., Raynal, J., Foli, A. & Pappalardo, G. (1972). *Nucl. Phys.*, **A189**, 353.

Beckerman, M., Salomaa, M., Sperduto, A., Enge, H., Ball, J., Di Rienzo, A., Gazes, S., Yan Chen, Molitoris, J. D. & Mao Nai-feng. (1980). *Phys. Rev. Lett*, **45**, 1472.

Beckerman, M., Salomaa, M., Sperduto, A., Molitoris, J. D. & Di Rienzo, A. (1983). *Phys. Rev.*, **C25**, 837.

Berry, M. V. (1966). *Proc. Phys. Soc.*, **89**, 479.

Berry, M. V. (1969). *J. Phys.*, **B2**, 381.

Berry, M. V. & Mount, K. E. (1972). *Rep. Prog. Phys.*, **35**, 315.

Berry, M. V. (1976). *Adv. in Phys.*, **25**, 1.

Bertsch, G., Borysowicz, J., McManus, H. & Love, W. G. (1977). *Nucl. Phys.*, **A284**, 399.

Blair, J. S. (1954). *Phys. Rev.*, **95**, 1218.

Blair, J. S. (1959). *Phys. Rev.*, **115**, 928.

Blocki, J., Randrup, J., Swiatecki, W. J. & Tsang, C. F. (1977). *Ann. Phys.*, **105**, 427.

Bohr, A. (1959). *Nucl. Phys.* **10**, 486.

Bonaccorso, A. & Brink, D. M. (1982). *Nucl. Phys.*, **A384**, 161.
Bond, P. D., Chasman, C., Le Vine, M.J., Schwartzschild, A. Z., Thorn, C. E. & Baltz, A. J. (1974), *Phys. Rev.*, **C9**, 2001.
Branden, M. E. (1982). *Phys. Rev. Lett.*, **49**, 1132.
Brillouin, L. (1926). *Compt. Rend.*, **183**, 24.
Brink, D. M. & Satchler, G. R. (1968). *Angular Momentum*, Oxford: OUP.
Brink, D. M. & Takigawa, N. (1977). *Nucl. Phys.*, **A279**, 159.
Brink, D. M., Grabowski, J. & Vogt, E. (1978). *Nucl. Phys.*, **A309**, 359.
Brink, D. M. & Satchler, G. R. (1981). *J. Phys.*, **G7**, 43.
Brink, D. M., Nemes, M. C. & Vautherin, D. (1983). *Ann. Phys.*, **147**, 171.
Brink, D. M. & Smilansky, U. (1983). *Nucl. Phys.*, **A405**, 301.
Broglia, R. A. & Winther, A. (1972). *Phys. Reports*, **4C**, 153.
Broglia, R. A., Dasso, C. H. & Winther, A. (1974*b*). *Phys. Lett.*, **53B**, 301.
Broglia, R. A., Landowne, S., Malfliet, R. A., Rostokin, V. & Winther, A. (1974*a*). *Phys. Reports*, **11C**, 1.
Broglia, R. A. & Winther, A. (1981). *Heavy-Ion Reactions*, Reading: Mass: Benjamin.
Broglia, R. A. (1983). *Nucl. Phys.*, **A409**, 163.
Broglia, R. A., Dasso, C. H., Landowne, S. and Winther, A. (1983). *Phys. Rev.*, **C27**, 2433.
Bryant, H. C. & Jarmie, N. (1968). *Ann. Phys.* **47**, 127.
Buck, B. & Pilt, A. A. (1977). *Nucl. Phys.* **A280**, 133.
Carlson, B. V. & Hussein, M. S. (1982). *Phys. Rev.*, **C26**, 2007.
Chester, C., Friedman, B. & Ursell, F. (1957). *Phil. Soc.*, **53**, 599.
Christensen, P. R. & Winther, A. (1976). *Phys. Lett.*, **65B**, 19.
Christensen, P. R. & Switkowski, Z. E. (1977). *Nucl. Phys.*, **A280**, 205.
Coleman, S. (1977). *Phys. Rev.*, **D15**, 2929.
Connor, J. N. L. (1973). *Molec. Phys.*, **26**, 1217.
Connor, J. N. L. (1974). *Molec. Phys.*, **27**, 853.
Connor, J. N. L. (1976). *Molec. Phys.*, **31**, 330.
Cook, J., Gils, H. J., Rebel, H., Majaka, Z. & Klewe-Nebenius, H. (1982). *Nucl. Phys.*, **A388**, 153, 173.
Cramer, J. G., de Vries, R. M., Goldberg, D. A., Zisman, M. S. & Maguire, C. F. (1976). *Phys. Rev.*, **C14**, 2158.
Crowley, B. J. B. (1978). *J. Phys.*, **A11**, 509.
Crowley, B. J. B. (1980). *Phys. Rep.* **57**, 48.
Cujec, B. & Barnes, C. A. (1976). *Nucl. Phys*, **A266**, 461.
Dar, A. (1966). *Nucl. Phys.*, **82**, 354.
da Silveira, R. (1973). *Phys. Lett.*, **45B**, 211.
da Silveira, R. & Leclercq-Willain, C. (1984). *Z. Phys.*, **A314**, 63.
Dasso, C. H., Landowne, S. & Winther, A. (1983*a*). *Nucl. Phys.*, **A405**, 381.
Dasso, C. H., Landowne, S. & Winther, A. (1983*b*). *Nucl. Phys.*, **A407**, 221.
Davies, H. (1963). *Proc. Camb. Phil. Soc.*, **59**, 147.
Davison, B. (1954). *Proc. Roy. Soc*, **A225**, 252.
Dean, D. R. & Rowley, N. (1984). *J. Phys.*, **G10**, 493.
de Boer, J., Dannhäuser, A., Massmann, H., Rösel, F. & Winther, A. (1977). *J. Phys.*, **G3**, 889.

Debye, P. J. (1908). *Physik Z.* **9**, 775.
Delbar, Th., Grégoire, Gh., Paic, G., Ceuleneer, R., Michel, F., Vanderpoorten, R., Budzanowski, A., Dabrowski, H., Freindl, L., Grotowski, K., Micek, S., Planeta, R., Strzalkowski, A. & Eberhard, K. A. (1978). *Phys. Rev.*, **C18**, 1237.
de Vries, R. M. & Clover, M. R. (1975). *Nucl. Phys.*, **A243**, 528.
Dingle, R. B. (1956). *Appl. Sci. Res.*, **B5**, 345.
Dingle, R. B. (1973). *Asymptotic Expansions.* London: Academic Press.
Di Salvo, E. & Viano, G. A. (1977). *Nuovo Cimento*, **A42**, 49.
Dollard, J. D. (1964). *J. Math. Phys.*, **5**, 729.
Dos Aidos, F. & Brink, D. M. (1985). *J. Phys.*, **G11**, 249.
Dos Aidos, F., Sukumar, C. V. & Brink, D. M. (1984). To be published.
Duistermaat, J. J. (1974). *Communs. Pure Appl. Maths.*, **27**, 207.
Ericson. T. E. O. (1966). *Preludes in Theoretical Physics.* ed. de Shalit, Feshbach & Van Hove, p. 321.
Fernandez, B. & Blair, J. S. (1969). *Phys. Rev.*, **C1**, 523.
Feshbach, H. (1958). *Ann. Phys.*, **5**, 357.
Feynman, R. P. (1948). *Rev. Mod. Phys.*, **20**, 367.
Feynman, R. P. & Hibbs, A. R. (1965). *Quantum Mechanics and Path Integrals.* New York: McGraw Hill.
Fiedeldey, H. (1967). *Nucl. Phys.*, **A96**, 463.
Ford, K. W. & Wheeler, J. A. (1959). *Ann. Phys.*, **7**, 259.
Frahn, W. E. & Venter, R. H. (1963). *Ann. Phys.*, **24**, 243.
Frahn, W. E. & Venter, R. H. (1964). *Ann. Phys.*, **27**, 135.
Frahn, W. E. (1966). *Nucl. Phys.*, **75**, 577.
Frahn, W. E. (1976). *Nucl. Phys.*, **A272**, 413.
Frahn, W. E. & Gross, D. H. E. (1976). *Ann. Phys.*, **101**, 520.
Frahn, W. E. & Rehm, K. E. (1978). *Phys. Rep*, **C37**, 1.
Frahn, W. E. (1980*a*). *Nucl. Phys.*, **A337**, 324.
Frahn, W. E. (1980*b*). *Phys. Rev.*, **C21**, 1820.
Frahn, W. E. & Hussein, M. S. (1980). *Nucl. Phys.*, **A346**, 237.
Frahn, W. E. (1984). *Heavy-Ion Science*, Vol. 1, ed. D. A. Bromley, New York: Plenum Press.
Freed, K. F. (1972). *J. Chem. Phys.*, **56**, 692.
Friedman, W. A. & Goebel, C. J. (1977). *Ann. Phys.*, **104**, 145.
Fuller, R. C. (1973). *Nucl. Phys.*, **A216**, 199.
Fuller, R. C. (1975). *Phys. Rev.*, **C12**, 1561.
Fuller, R. C. & Moffa, P. J. (1977). *Phys. Rev.*, **C15**, 266.
Furry, W. H. (1947). *Phys. Rev.*, **71**, 360.
Garrod, G. (1966). *Rev. Mod. Phys.*, **38**, 483.
Goldberg, D. A., Smith, S. M. & Burdzik, G. F. (1974). *Phys. Rev.*, **C10**, 1362.
Goldberg, D. A. & Smith, S. M. (1974). *Phys. Rev.*, **33**, 715.
Goldberg, D. A. (1975). *Phys. Lett.*, **55B**, 59.
Gubler, H. P., Plattner, G. R., Sick, I., Traber, A. & Weiss, W. (1977). *Nucl. Phys.*, **A284**, 114.
Gubkin, I. A. (1978). *Sov. J. Nucl. Phys.*, **28**, 338.
Gutzwiller, M. C. (1967). *J. Math. Phys.*, **8**, 1929.
Gutzwiller, M. C. (1969). *J. Math. Phys.*, **10**, 1004.

Gutzwiller, M. C. (1970). *J. Math. Phys.*, **11**, 1791.
Hahne, F. J. W. (1967). *Nucl. Phys.*, **A104**, 545.
Hansen, O., Videbeck, F., Flynn, E. R., Peng, J. C. & Cizeweski, J. A. (1981). *Nucl. Phys.*, **A364**, 144.
Hasan, H. & Brink, D. M. (1978). *J. Phys.*, **G4**, 1573.
Hasan, H. & Brink, D. M. (1979). *J. Phys.*, **G5**, 771.
Hasan, H. & Warke, C. S. (1981). *J. Phys.*, **G7**, 1501.
Hiebert, J. C. & Garvey, G. T. (1964). *Phys. Rev.*, **135B**, 346.
Hill, D. L. & Wheeler, J. A. (1953). *Phys. Rev.*, **89**, 1102.
Horiuchi, H. (1980*a*). *Prog. Theoret. Phys.*, **63**, 725.
Horiuchi, H. (1980*b*). *Prog. Theoret. Phys.*, **64**, 184.
Horiuchi, H. (1983). *Prog. Theoret. Phys.*, **69**, 886.
Hussein, M. S. & McVoy, K. W. (1984). To be published.
Igo, G. (1959). *Phys. Rev.*, **115**, 1665.
Jacob, P. M. & Smilansky, U. (1983). *Phys. Lett.*, **127B**, 313.
Jeffreys, H. (1923). *Proc. London Math. Soc.*, **2**, 23, 428.
Jeffreys, H. & Jeffreys, B. S. (1956). *Methods of Mathematical Physics*, Sections 17, 121, 17.13. Cambridge University Press.
Kahana, S. & Baltz, A. J. (1977). *Adv. in Nucl. Phys.*, **9**, 1.
Kauffman, S. K. (1977). *Z. Phys.*, **A282**, 163.
Keller, J. B. (1955). *J. App. Phys.*, **26**, 961.
Keller, J. B. (1957). *J. App. Phys.*, **28**, 426.
Keller, J. B. (1958). *Ann. Phys.*, **4**, 180.
Keller, J. B. (1962). *J. Opt. Soc.*, **52**, 116.
Kemble, E. C. (1935). *Phys. Rev.*, **48**, 549.
Knoll, J. & Schaeffer, R. (1976). *Ann. Phys.*, **97**, 307.
Knoll, J. & Schaeffer, R. (1977). *Phys. Rep.*, **C31**, 159.
Kobono, S., Bond, P. D., Horn, D., & Kenner, T. R. (1980). *Phys. Rev.*, **C21**, 459.
Kobos, A. & Satchler, G. R. (1984). *Phys. Rev.*, **C30**, 403.
Koeling, T. & Malfliet, R. A. (1975). *Phys. Rep.*, **C22**, 182.
Kramers, H. A. (1926). *Z.* Physik, **39**, 828.
Landau, L. D. & Lifshitz, E. M. (1958). *Quantum Mechanics*, London: Pergamon Press.
Landowne, S., Dasso, C. H., Nilsson, B. S., Broglia, R. A. & Winther, A. (1976). *Nucl. Phys.*, **A259**, 99.
Landowne, S. (1979). *Phys. Rev. Lett.*, **42**, 633.
Landowne, S. & Wolter, H. H. (1981). *Nucl. Phys.*, **A351**, 171.
Langer, R. E. (1934). *Bull. Amer. Math. Soc.*, **40**, 545.
Langer, R. E. (1937). *Phys. Rev.*, **51**, 669.
Lee, S. Y. & Takigawa, N. (1978). *Nucl. Phys.*, **A308**, 189.
Lee, S. Y., Takigawa, N. & Marty, C. (1978). *Nucl. Phys.*, **A308**, 161.
Lee, S. Y. & Takigawa, N. (1983). *Phys. Rev.*, **C28**, 1123.
LeVine, M. J., Baltz, A. J., Bond, P. D., Garrett, J. D., Kahana, S. & Thorn, C. E. (1974). *Phys. Rev.*, **C10**, 1602.
Levit, S. & Smilansky, U. (1977*a*). *Ann. Phys.*, **103**, 198.
Levit, S. & Smilansky, U. (1977*b*). *Ann. Phys.*, **108**, 165.

Levit, S. (1978). *Path Integrals and Semi-classical Methods for Nuclear Heavy-Ion Reactions.* PhD Thesis. Rehovot: the Weizmann Institute of Science.
Levit, S., Möhring, K., Smilansky, U. & Dreyfus, T. (1978). *Ann. Phys.*, **114**, 223.
Levy, B. R. & Keller, J. B. (1959). *Commun. Pure App. Math.*, **12**, 159.
Lindsay, R. & Rowley, N. (1984). *J. Phys.*, **G10**, 805.
Love, W. G., Terasawa, T. & Satchler, G. R. (1977). *Nucl. Phys.*, **A291**, 183.
McVoy, K. W. (1978). *Notas de Fisica*, vol. 1, no. 4. Mexico: Univ. of Mexico.
Malfliet, R. A., Landowne, S. & Rostokin, V. (1973). *Phys. Lett.*, **44B**, 238.
Marcus, R. A. (1971). *J. Chem. Phys.*, **54**, 3965.
Maslov, V. P. (1972). *Théorie des Perturbations et Méthodes Asymptotiques.* Paris: Dunod.
Messiah, A. (1959). *Méchanique Quantique.* Paris: Dunod.
Miller, S. E. & Good, R. H. (1953). *Phys. Rev.*, **91**, 174.
Miller, W. H. (1968*a*). *J. Chem. Phys.*, **48**, 464.
Miller, W. H. (1968*b*). *J. Chem. Phys.*, **48**, 1651.
Miller, W. H. (1969). *J. Chem. Phys.*, **51**, 3631.
Miller, W. H. (1970). *J. Chem. Phys.*, **53**, 1949.
Miller, W. H. (1974). *Adv. Chem. Phys.*, **25**, 69.
Miller, W. H. (1975). *Adv. Chem. Phys.*, **30**, 77.
Michel, F. & Vanderpoorten, A. (1977). *Phys. Rev.*, **C16**, 142.
Michel, F., Albinski, J., Belevy, P., Delbar, Th., Grégoire, Gh., Tasiaux, B. & Reidemeister, G. (1984). Preprint.
Migdal, A. B. (1977). *Qualitative Methods in Quantum Theory.* Reading, Mass.: W. A. Benjamin, Inc.
Möhring, K., Levit, S. & Smilansky, U. (1980). *Ann. Phys.*, **127**, 198.
Morse, M. (1934). *The Calculus of Variations in the Large.* Providence, Rhode Island: Amer. Math. Soc.
Morse, M. (1973). *Variational Analysis.* New York: Wiley.
Mott, N. F. (1931). *Proc. Camb. Phil. Soc.*, **27**, 553.
Mott, N. F. & Massey, H. S. W. (1949). *Theory of Atomic Collisions*, 2nd Ed. Oxford: OUP.
Nussenzweig, H. M. (1965). *Ann. Phys.*, **34**, 23.
Nussenzweig, H. M. (1969). *J. Math. Phys.*, **10**, 82.
Nussenzweig, H. M. (1977). *Sci. Am.*, **236**, 116.
Nussenzweig, H. M. (1979). *J. Opt. Soc. Am.*, **69**, 1068.
Paschopoulos, I., Fisher, P. S., Jelley, N. A., Kahana, S., Pilt, A. A., Rae, W. D. M. & Sinclair, D. (1975). *Nucl. Phys.*, **A252**, 173.
Pechukas, P. (1969). *Phys. Rev.*, **181**, 166, 174.
Perey, F. G. & Buck, B. (1962) *Nucl. Phys.*, **32**, 353.
Perey, F. G. (1963). *Direct Interaction and Nuclear Reaction Mechanism*, eds. E. Clementel & C. Villi. New York: Gordon & Breach Inc.
Potgieter, J. M. & Frahn, W. E. (1967). *Nucl. Phys.*, **A92**, 84.
Reisdorf, W., Hessberger, F. P., Hildebrand, K., Hofmann, S., Münzerberg, G., Schmidt, K. H., Schneider, J. H., Schneider, W. F. W., Sümmerer, K., Wirth, Kratz, J. V. & Schlitt, K. (1982). *Phys. Rev. Lett.*, **49**, 1811.
Reeves, H. (1966). *Astrophys. J.*, **146**, 447.

Rowley, N. (1974). *Nucl. Phys.*, **A219**, 93.

Rowley, N., & Plagnol, E. (1975). *Phys. Lett.*, **56B**, 221.

Rowley, N. & Marty, C. (1976*a*). *Nucl. Phys.*, **A266**, 494.

Rowley, N. & Marty, C. (1976*b*). *J. Phys.*, **G2**, 217.

Rowley, N., Doubre, H. & Marty, C. (1977). *Phys. Lett.*, **69B**, 149.

Satchler, G. R. (1983*a*). *Direct Nuclear Reactions*. Oxford: OUP.

Satchler, G. R. (1983*b*). *Nucl. Phys.*, **A409**, 3.

Satchler, G. R. & Love, W. G. (1979). *Phys. Rep.*, **55**, 183.

Satchler, G. R., Fulmer, C. B., Auble, R. L., Ball, J. B., Bertrand, F. E., Erb, K A. & Hensley, D. C. (1983).*Phys. Lett.*, **128B**, 147.

Schiller, R. (1962). *Phys. Rev.*, **125**, 1109.

Schröder, W. U. & Huizenga, J. R. (1977). *Ann. Rev. Nucl. Sci.*, **27**, 465.

Siemens, P. J. & Becchetti, F. D. (1972). *Phys. Lett.*, **42B**, 389.

Schulman, L. S. (1981). *Techniques and Applications of Path Integration*. New York: John Wiley.

Stokes, G. G. (1904). *Maths and Phys. Papers*. London: Cambridge University Press.

Stokstad, R. G., Eisen, Y., Kaplanis, S., Pelte, D. & Smilansky, U. & Tserruya, I. (1978). *Phys. Rev. Lett.*, **41**, 465.

Strutinsky, V. M. (1964). *JETP*, **19**, 1401.

Strutinsky, V. M. (1973). *Phys. Lett.*, **44B**, 245.

Sukumar, C. V. (1984). *J. Phys.*, **G10**, 81.

Sukumar, C. V. & Brink, D. M. (1983). *Nucl. Phys.*, **A404**, 121.

Szegö, G. (1934). *Proc. Lon. Math. Soc.*, **2–36**, 427.

Takemasa, T. & Tamura, T. (1978). *Phys. Rev.*, **C18**, 1282.

Takigawa, N. & Lee, S. Y. (1977). *Nucl. Phys.*, **A292**, 173.

Takigawa, N. & Bertsch, G. (1984). To be published.

Taylor, J. R. (1972). *Scattering Theory*. New York: John Wiley.

Thom, R. (1975). *Structural Stability and Morphogenesis*. Reading, Mass.: Benjamin.

Traber, A., Trautmann, D. & Rösel, F. (1977). *Nucl. Phys.*, **A291**, 221.

van Horn, H. M. & Salpeter, E. E. (1967). *Phys. Rev.*, **157**, 751.

van Vleck, J. H. (1928). *Proc. Nat. Acad. Sci.*, **14**, 178.

Vaz, L. C., Alexander, J. M. & Satchler, G. R. (1981). *Phys. Rep.*, **69C**, 373.

Venter, R. H. (1963). *Ann. Phys.*, **25**, 405.

Venter, R. H. & Frahn, W. E. (1964). *Ann. Phys.*, **27**, 401.

Vinh Mau, N. (1977). *Phys. Lett.*, **71B**, 5.

von Oertzen, W. (1970). *Nucl. Phys.*, **A148**, 529.

Wentzel, G. (1926). *Z. Physik*, **38**, 518.

Wong, C. Y. (1973). *Phys. Rev. Lett.*, **31**, 766.

Yennie, D. R., Ravenhall, D. G. & Wilson, R. N. (1954). *Phys. Rev.*, **95**, 500.

# Index